MW00442009

MINING HISTORY
AND GEOLOGY
OF
JOSHUA TREE
NATIONAL PARK

Photo by Peter S. Gorwin, 2004

Dedicated to all those who draw scientific and artistic inspiration from Joshua Tree National Park, and to the spouses and loved ones who support them.

A special thank you to my buddies and fellow Desert Rats, for helping me discover the history and geology of California.

MRE

MINING HISTORY AND GEOLOGY

OF

JOSHUA TREE NATIONAL PARK

San Bernardino and Riverside Counties, California

Editor
Margaret R. Eggers, PhD

San Diego Association of Geologists
c/o Sunbelt Publications
P.O. Box 191126
San Diego, CA 92159-1126
meggers@eggersenv.com

SAN DIEGO ASSOCIATION OF GEOLOGISTS

Mining History and Geology of Joshua Tree National Park

Copyright © 2004 by the San Diego Association of Geologists (SDAG)
Individual papers, photos, and artwork copyrighted by the authors and used with permission
All rights reserved. First edition 2004, sixth printing 2015

Edited and designed by Margaret R. Eggers
Cover design by Margaret R. Eggers,
and Phil Farquharson, CG-Squared Productions
Printed in the United States of America

Distributed by Sunbelt Publications
www.sunbeltpublications.com
Please direct comments and inquiries to:
SDAG, P.O. Box 191126, San Diego, CA 92159-1126
www.sandiegogeologists.org

18 17 16 15   9 8 7 6

Library of Congress Cataloging-in-Publication Data

Mining history and geology of Joshua Tree National Park :
San Bernardino and Riverside counties, California / editor,
Margaret R. Eggers.-- 1st ed.
    p. cm.
  Includes bibliographical references.
  ISBN 0-916251-70-5
  1. Mines and mineral resources--California--Joshua Tree
National Park--History. 2. Mineral industries--California--
Joshua Tree National Park--History. 3. Geology--
California--Joshua Tree National Park. I. Eggers,
Margaret R., 1959-
  TN24.C2M49 2004
  553'.09794'97--dc22

                          2004023228

Front Cover art: Untitled Joshua Tree Scene by Tom Cranham (1934-1997). Cranham was a special effects illustrator and artist for numerous feature films and TV series. He also served as vice president of the Motion Picture Illustrator and Matte Artists Guild #790 for 18 years. In his off hours, he painted many canvases of the California landscape and Missions. Reproduced here courtesy Greg T. Cranham, PG, CEG.

Back Cover photo "*Joshua Tree Shadow*", 2004,
by John D. Clark.
Cover background photo "*Joshua Tree Starburst*", 2004,
by John D. Clark, www.johndclark.com.

Inside front cover map "*Geological Map of the North Central Portion of Joshua Tree National Park*", prepared by D. D. Trent.

Inside back cover *Two Views of Joshua Tree National Park from Space*, Prepared by Phil Farquharson, CG-Squared Productions (www.cg-squared.com).

All other photos and illustrations copyrighted by respective authors unless otherwise noted, and used here with permission.

APRG FG #47

Photo by Phil Farquharson
Taken at Barker Dam, September 2004

"...but there were mountain sheep, and we would see them quite often. They used to come and stand on the rocks once in a while. You'd see a big old buck standing up there especially if the mill was running or some activity or noise was going on at the ranch. They'd sit there and stare for a half an hour. You know, watching. And when we used to run the old two-stamp mill over at the Wall Street Mill site, they used to come and stand up there on the rocks. There might be four or five of them at a time and they'd watch. They were beautiful."

Recollection of Willis Keys. From "Growing up at the Desert Queen Ranch",1997, by Willis Keys and Art Kidwell, published by the Joshua Tree Park Association.

CONTENTS

SAN DIEGO ASSOCIATION OF GEOLOGISTS

ANNUAL FIELD TRIP
OCTOBER 2004

SDAG 2004 OFFICERS

President
Monte Murbach
Petra Geotechnical

Vice-President
Margaret R. Eggers
Eggers Environmental, Inc.

Secretary
David Bloom
Anteon Corporation

Treasurer
Phil Farquharson
CG-Squared Productions

SAN DIEGO ASSOCIATION OF GEOLOGISTS

SAN DIEGO ASSOCIATION OF GEOLOGISTS

2004 CORPORATE SPONSORS

Dr. Pat Abbott

Dr. Richard Berry

Sherry Bloom,
Coldwell Banker Residential Real Estate

Steven N. Bradley,
Testing Engineers - San Diego, Inc.

Curtis R. Burdett,
Christian Wheeler Engineering

Pat Brooks and Julie Crosby
Construction Testing & Engineering, Inc.

Joe Corones, City of San Diego

Robert Crisman, Geo Soils, Inc.

Dr. Margaret R. Eggers, PG, CHG,
Eggers Environmental, Inc.

William J. Elliott, Engineering Geologist

Phil Farquharson, CG-Squared Productions

Katherine Freese

Fugro West, Inc.

GeoPacifica, Inc., James Knowlton

Carolyn Glockhoff, Caro-Lion Enterprises

Jonathan Goodmacher, PSI

Dr. Sarah Gray, University of San Diego

Hargis + Associates, Inc.

Rob Hawk, City of San Diego

John Hoobs, Geocon Inc.

Kleinfelder

Lowell Lindsay, Sunbelt Publications

SAN DIEGO ASSOCIATION OF GEOLOGISTS

2004 CORPORATE SPONSORS
(continued)

Dr. Monte Marshall

Laura Maghsoudlou, EnviroMINE, Inc.

Ninyo & Moore

**John Peterson,
Peterson Environmental Services**

Bob Smillie, TerraCosta Consulting Group

Donald Sorben

David and Jan Steller

Dr. Anne Sturz

**Sue Tanges, Southland Geotechnical
Consultants**

**Malcolm Vinje,
Vinje & Middleton Engineering, Inc.**

Carole L. Ziegler

The generous support of our corporate sponsors helps the SDAG provide annual scholarships to undergraduate and graduate students in the geological sciences; provides for reduced fees for student attendance to meetings; provides support for a monthly newsletter; the annual field trip and production of the annual SDAG field trip guidebook. To contact any of our generous sponsors, please visit the SDAG website at www.sandiegogeologists.org, where many of our sponsors have direct links. To become a sponsor of SDAG, contact any of our SDAG officers.

FOREWORD

I became a college teacher in the early 1960s after having worked in the oil patch for several years. Initially, my geology field trips involved excursions to those areas that dealt with geology that I knew well: sedimentary rocks, structural geology and fossils. The eastern Mojave Desert provided the perfect field area for such trips and Joshua Tree National Monument (now upgraded to Park status) offered a suitable camp site along the way. We could leave campus on a Friday afternoon, camp there that night, and depart early on Saturday allowing us a full unhurried day in the field looking at the really important stuff in the Marble and Providence Mountains. Of course, before leaving Joshua Tree, we'd spend an hour or so discussing plutons, their emplacement, and a bit about metamorphic rocks -- after all, this kind of terrane deserved some attention, but petroleum isn't associated with such crystalline rocks so they weren't important enough to devote significant time looking at such features. However, the students questioned why the bold rock outcroppings at Joshua Tree looked the way they do, and the answer I had for them wasn't entirely satisfactory. I'd been taught that such landscapes were typical of those that had formed by weathering and erosion under arid conditions. Their question, and my dissatisfaction with what I'd been taught, nagged at me. And this illustrates an important point: If you can't learn from your students, they're not worth much. Thus began my love affair with the Joshua Tree landscape.

It didn't take long before an entire weekend field trip became devoted to examining the landscape and petrology of Joshua Tree. It became obvious that here, less than a three hour drive from some 15 million people in the Los Angeles metropolitan area, was magnificent scenery that exposed a wonderful story that included a record of how continents grow, some of the oldest rocks in California, superb exposures of granitic intrusives, active

(photo by Peter S. Gorwin, 2004)

faults and their attendant fault scarps, mining geology, a large, plateau-like "fossil" landscape caught between two major faults, columnar volcanics, and a variety of arid region geomorphic features. Added delights included the artifacts from the early ranching and gold mining days of the 19th century, Indian petroglyphs and pictographs, and gorgeous displays of spring wild flowers. But still, I lacked a solid explanation of the park's geomorphology until I happened upon Arthur Holmes' (1965) explanation for the origin of the granite tors of Dartmoor. His explanation seemed clearly applicable and I shamelessly adopted it to explain the inselbergs and attendant features of the park. Then, in 1972, Ted Oberlander explained the origin of similar regions in the Mojave, which confirmed my hypothesis, and tied everything together to explain the Joshua Tree landscape. It was now obvious that the landscape was not typical of arid region processes at all, but was the product of several million years of climate variation tied to global climate change dating back perhaps to the Miocene – and this long before global climate change had become the buzz word we hear tossed about so much today.

Visiting Joshua Tree National Park is, in a way, encountering our roots, the Park serving as a looking glass into the Earth's past history. The area's botany and the two plant communities (Colorado Desert and Great Basin-Mojave Desert) were the initial stimuli responsible for setting aside the region for protection by the National Park Service, but the geologic features could have served equally well. Consequently, its Park status has resulted in minimal development and protection of the area allowing visitors a fine opportunity to grasp our place in the grand scheme of things and a chance to perceive our sense of place and time in a setting isolated from the congestion of an increasingly frenetic urban world.

Dee Trent
Claremont, California

ACKNOWLEDGEMENTS

Over the past years, I have been lucky to travel around southern California with many fellow geologists who really helped me -- a transplanted southern girl with a love of the Smoky Mountains -- learn to appreciate the very different geology and landscape of California. But although I had made many trips to Death Valley, Yosemite, and the Mammoth and Mono areas I had not yet been to Joshua Tree. When it became my turn to organize the SDAG annual field trip, I wanted to focus on somewhere I had not been before. Although Joshua Tree was so close, I had never investigated what it had to offer. So three years ago I started making treks with husband and dog in tow.

The result is this field guide. Like most endeavors, it took a lot more effort and time than was planned. But the terrific contributions of many generous folks has produced what I think is a truly unique, useful and informative companion for a Joshua Tree National Park visit.

I would like to thank the following people for their help, insight, time and effort that made this publication a success:

- Dr. Andy Barth at Purdue summarized his current research on the age dating and timing of events in the Park especially for this volume. Andy's work represents the latest in our understanding of how the rocks and terrane of Joshua Tree formed.

- Dr. John D. Clark, my talented husband, has provided this publication with many of his dramatic large format black & white photographs – including the beautiful back cover shot of the Joshua Tree shadow. John accompanied me on many a recon trip with patience and good humor.

- Greg Cranham, past SDAG president, friend, colleague, and fellow Desert Rat graciously allowed us to photograph and reproduce his father's wonderful painting of a Joshua Tree scene for our cover.

- Phil Farquharson of CG-Squared Productions graciously did all the final layout and polishing of our covers. Phil also helped me with field reconnaissance and GPS tracking for the road logs, provided photos (Bighorns at Barker Dam!) and all-round support.

- Peter Gorwin (www.pgphotographics.com) and his wife Jane accompanied us on reconnaissance visits and Peter contributed many of his beautiful Joshua Tree photographs to accent our guidebook.

- Dr. Rick Hazlett of Pomona College provided technical review to Andy Barth's contribution.

- Woody Higdon of Geo-Tech-Imagery (geo-tech-imagery.com) prepared aerial photograph mosaics of areas of interest.

- Dr. David Kimbrough of San Diego State University provided technical review for Ms. Probst's article on her study of Joshua Tree basalts.

- Lowell Lindsay, owner - with his wife Diana - of Sunbelt Publications and fellow SDAG member and past SDAG President himself, has been instrumental in helping this first-time editor/author nurse this project to the reality of a printing press.

- Michael Palmer, longtime colleague, friend and fellow Desert Rat, provided his editorial review of the mining equipment article.

- Kelly Probst, an undergraduate student of Andy Barth's, prepared her article on Joshua Tree basalts just before moving to the University of Florida to pursue her graduate degree in geology. I think those 'gators are lucky to have her!

- My biggest, heartfelt THANK YOU goes to Dr. Dee Trent. It is no exaggeration that without him this volume would not have been impossible. Dee generously provided me with contacts for technical contributions, many graphics, photos, and guidance in compiling this guidebook. He prepared the article on the Park mine sites; inspired me and helped me to write the article on historic mining equipment, and worked with me to compile the article on the overview of park geology. His generosity, good humor, and decades of experience in the study of and writing about Joshua Tree geology were invaluable in assembling this volume.

Margaret R. Eggers
Oceanside, California

INTRODUCTION

Margaret R. Eggers, PhD
Eggers Environmental, Inc., Oceanside, California

Joshua Tree National Park has much to offer, not just for the geologist but also for the historian, archeologist, botanist and any type of naturalist. The landscape of this unique setting has also inspired artists to capture scenes of Joshua Tree in paint and on film. The cover of this guide and many of the photographs contained within these pages provide beautiful examples of the Park's artistic inspiration.

The nearly 800,000 acres within the Park boundaries encompass portions of both the Mojave and Colorado Deserts. These two large ecosystems, determined largely by elevation changes, meet about halfway through the Park. Below 3,000 feet in the eastern half of the park, the Colorado Desert is host to abundant creosote, ocotillo and cholla cactus. In the higher elevation Mojave Desert to the west, which is also cooler and wetter, one can find the Park's namesake. The Joshua Tree, synonymous with the Mojave Desert, grows in extensive stands throughout the western half of the Park, intermingling with the large granitic outcrops to produce the beautiful scenery characteristic of the Park.

In the 1800s, cattlemen migrated to this area looking for grazing territory. They built dams to increase the capacity of natural water catchments in the granite terrane which were already known to the local Indians and wildlife. Archeological evidence of the early inhabitants in the form of petroglyphs and stone grinding mortars in the granites, intermingle with later dam construction in the Barker Dam and Squaw Tank areas.

Miners also discovered gold throughout the region. Their activities left behind 288 abandoned mine sites within the Park and countless historic sites as a record of California's mining history. Papers on the Park's mines and historic ore processing equipment and techniques will help the visitor understand and enjoy the many mine sites scattered throughout the Park.

In the early 1900s, people became concerned with the rapid decline in the ecological health of the region. Human activity was leaving more and more scars on the desert. Suburban landscapers treated the area like a public nursery, decimating many stands of cacti, and Joshua Trees were burned and vandalized indiscriminately. Thanks to the actions of dedicated individuals, such as Minerva Hamilton Hoyt, the Park has been preserved for all of us to enjoy. The story of Mrs. Hoyt's efforts to preserve the Joshua Tree area is told herein by Bob Cates. Clearly, one individual CAN make a difference. This lesson is as important for us today as it was a century ago.

While the Joshua Tree area initially gained protection as a National Monument in 1936, many subsequent efforts on the part of conservationists continued to expand and protect this unique desert area. The eastern portion of the Oasis of Mara was added to the Park in 1950. In 1976, Congress designated much of the Monument area as "wilderness," meaning mechanized equipment, development or human occupation is not allowed. These wilderness areas offer the increasingly rare opportunity to experience the solitude and quiet of an unconfined, primitive desert experience. Most importantly, in 1994 President Bill Clinton signed the Desert Protection Bill, elevating the original Monument to National Park status, and adding 234,000 acres to the Park. These additional acres were especially important because they modified the Park boundaries to include natural features and complete ecological units which encompassed entire watersheds and range areas for the Desert Big Horn sheep.

For the "rock-hound" or geologist, Joshua Tree National Park offers a wealth of rock types and landforms to study and enjoy. This guide contains an overview of the geology in the park, as well as articles by Professor Andy Barth and Kelly Probst which represent the latest research and understanding of the chronology and genesis of the faults and rock units within the Park. A review of the reference sections of these articles will lead you to more detailed maps, papers and books to expand your understanding. One thing I have observed during my visits to Joshua Tree is that there's always another trail, another outcrop, another mine, another area to explore. I am certain that your experiences here will bring you back to experience and learn more. Enjoy!

MINERVA HAMILTON HOYT AND THE PRESERVATION OF JOSHUA TREE

"Mrs. Hoyt's Monument"
Excerpted with permission of the author from
"Joshua Tree National Park: A Visitor's Guide,"
by **Robert B. Cates**, 1995, Live Oak Press, Chatsworth, CA

(Photo by Peter Gorwin, 2004)

Minerva Hamilton Hoyt was a large and stately woman, aristocratic and wealthy, and deeply involved with all the socially correct charitable organizations of 1900s Los Angeles and Pasadena. She was what one might refer to as a "society matron." She most certainly was not a sun-hardened desert rat. But something in the open spaces and comforting solitude of the California desert stirred her soul, with the result that she almost single-handedly led the crusade to create Joshua Tree National Monument, the predecessor to the National Park.

Joshua Tree is just one of a multitude of superb parks and monuments that we Americans tend to take for granted. But parks do not simply spring into existence, and in taking them for granted, we ignore the efforts and sacrifices of those who labor so diligently to create these havens of wild beauty. The story of Minerva Hoyt and her legacy deserves telling not just for history's sake, but as an example of how conservationists have always and still continue to function – largely upon the storehouse of their own hopes and inspiration.

At the beginning of the twentieth century the California desert was a place to be avoided unless you were a hermit or one of the few hardy miners bent on hunting and digging for mineral wealth. The popular conception of the desert was that it was always hot, uncomfortable, and dangerous – and it could indeed be all these things. Travel to the desert and across its vast expanses was a difficult undertaking. But by the 1920s a change was taking place, a change that created far-reaching problems for the desert that have yet to be satisfactorily resolved.

It was the automobile revolution, and with it came the discovery that the wagon roads developed by teamsters hauling freight to isolated desert sites could be negotiated by the tin lizzie. Ever-increasing numbers of tourists came pouring out of the coastal metropolitan areas of Southern California into a new-found playground where the air was pure and the sky nearly always clear.

What began to trouble Mrs. Hoyt and others was not the mere presence of greater numbers of people on the desert – its vastness was capable of absorbing the new multitudes unnoticed – but the increasing incidence of vandalism in the form of wholesale uprooting of native desert vegetation to be transplanted to city dwellers' gardens. With the recreational discovery of the desert during the 1920s, cactus gardens seemed to suddenly sprout around every home and bungalow in Southern California. As time went on, the problem grew to epidemic proportions. Fully-grown palm trees were being removed from native oases to grace urban patios. Far worse was the decimation of the Devil's Garden, the sparse remains of which are still visible along the first five miles of Highway 62 after it branches from Interstate 10 north of Palm Springs. Here once was located the epitome of Colorado Desert plant life, thousands of acres containing the most concentrated congregation of yuccas and cacti in all the California desert. George Wharton James described the site in his book *Wonders of the Colorado Desert* (1906): "When we find ourselves on the mesa, we begin to understand why this is called by the prospector 'the devil's garden.' It is simply a vast, native, forcing ground for a thousand varieties of cactus. They thrive here as if specially guarded...I know of no place where so many varieties are to be found as in this small area near the Morongo Pass." By 1930 it had been stripped bare, to the extent that even now, after decades of recovery, the passing motorist barely notices the sparse covering of cacti that has slowly reclaimed the area.

Just as exasperating was the vandalism directed toward the Joshua Tree. Travelers crossing the desert at night were setting fire to individual Joshua trees to guide other motorists. Then on June 14, 1930, the tallest Joshua tree ever known to exist was put to the torch and killed by unknown vandals.

The job of obtaining protection for the "worthless" desert was a long one (and in fact continues to this day). Mrs. Hoyt realized that her first task would be one of educating the public and government to appreciate desert values. In the late 1920s she initiated the process by mounting a series of highly acclaimed desert exhibitions, consisting of realistic habitat displays, in the United States and England, winning the support of influential scientists and the backing of the country's many garden clubs. In 1930 the next stage in her campaign evolved with her formation of the International Desert Conservation League, created "...to respond to an urgent demand for the protection of desert life and conservation of desert beauty spots."

Shortly thereafter Mrs. Hoyt first publicly broached the idea of a large federal desert park, to encompass over 1,000,000 acres located roughly where Joshua Tree National Monument came to be established, but stretching eastward all the way to the Colorado River. Horace M. Albright, then Director of the National Park Service, was sympathetic, but due to the pressing tasks facing his overworked agency in establishing two other desert projects (Saguaro National Monument in Arizona and Death Valley National Monument in California), he felt he could not commit Park Service resources for a third desert effort. Albright also recognized that a checkerboard pattern of land ownership within Mrs. Hoyt's proposed park boundaries (a Southern Pacific Railroad land grant consisting of every other section of land extended over much of the area), combined with a scattering of homesteads and mining claims, presented additional complications.

(Photo by Peter Gorwin, 2004)

Despite this disappointment, Mrs. Hoyt remained convinced that a federal desert park was required. In 1933 a bill was passed in the California legislature to create a "California Desert Park" that would have encompassed much of the area in Mrs. Hoyt's proposal. In response to this threat to her federal project, Mrs. Hoyt asked Governor Rolph to give her more time to plead her case before the federal government. Rolph cooperated, not only vetoing the state bill, but also presenting Mrs. Hoyt with a letter of introduction to newly-elected President Franklin D. Roosevelt. In the subsequent meeting with Roosevelt and Secretary of the Interior Harold Ickes she was buoyed by their receptivity to her proposal. Ickes summed it up: "The President is for this and I am for this."

Now Mrs. Hoyt drew upon the expertise of two eminent scientists, Drs. Philip Muna and Edmund C. Jaeger, to compile a two-volume photographic and written description of the outstanding plants, scenery, and recreational attractions of the area under proposal. This resource survey served as the basis for the establishment of Joshua Tree National Monument.

Neither Ickes nor Roosevelt was intimidated by the private property problems associated with the project. Four months after Hoyt's meeting with them, an area of 1,136,000 acres was withdrawn for consideration as a new national park or monument.

Success for Mrs. Hoyt? Not yet. During the next three years she tirelessly conducted official tours of the area, while simultaneously cajoling, badgering, and generally lobbying government officials to formally establish a national monument of maximum size. Finally, on August 10, 1936, President Roosevelt signed the proclamation setting aside 825,000 acres of prime California desert as Joshua Tree National Monument.

Minerva Hoyt, who has been referred to as the "Apostle of the Cacti" due to her unflagging devotion to desert preservation, died in 1945. Through her self-sacrifice, you and I and all the people who visit this part of California have the opportunity to scramble over the gigantic boulders in the Wonderland of Rocks, explore the scenic Joshua Tree forests, wonder at the vastness of Pinto Basin, or trek into the backcountry where the clever hand of man has not wreaked its typical havoc.

"Balancing," granitic monoliths, Hidden Valley (Photo by John D. Clark, 2004).

OVERVIEW OF THE LANDSCAPE AND GEOLOGY OF JOSHUA TREE NATIONAL PARK

Margaret R. Eggers, PhD, *Eggers Environmental, Inc.,*
Oceanside, CA 92054
D.D. "Dee" Trent, PhD, *Professor Emeritus,*
Citrus College, Glendora, CA 91741

INTRODUCTION

Traveling through Joshua Tree National Park, most people are struck by the dramatic and unique rock forms that sit on the landscape like monumental sculptures. While there are many landscapes which are also dominated by granitic outcrops, the look of the Joshua Tree landscape is unique and recognizable. But why do these rocks look so distinctive? It seems that a fortunate geological coincidence combining granitic rocks with the right weathering environment, climate change, and the tectonic and structural setting have produced this scenic and magical landscape.

Figure 1. Sculpted granitic forms at Jumbo Rocks (photo by John D. Clark, 2004).

A Landscape Controlled by Joints and Weathering

In some granitic outcrops, well developed joint sets produce large, rectangular blocks which then weather spheroidally in place (Figure 1). This may result in huge, granite "marbles" which appear to balance delicately on the tops of large outcrops. Conjugate joint sets are joint pairs set at 60- and 120-degrees from each other. These joints develop in response to a directional, compressive stress. If you look at the photo in Figure 1, the open end of the 60-degree angle joints are on the top and bottom, while the 120-degree angled joints are open to the left and right. This means that the orientation of the stress, relative to the <u>current</u> orientation of the outcrop, is compressional from the top and the bottom toward the center.

Other subtle variations in joint patterns and weathering produce more horizontal joints, which result in boulder piles (Figure 2). Differences in the intensity, orientation and frequency of joints result in the subtle variations which make each Joshua Tree boulder pile just a little bit different from all the rest. In some areas of the park, conjugate joint sets may not be strongly expressed or absent, and other types of joints control the landscape. A dominance of strong, vertical joints tends to produce tall, massive, granitic monoliths reminiscent of Stonehenge (Figure 3).

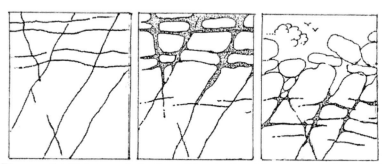

Figure 2. Origin of Joshua Tree's boulder piles: chemical weathering occurs along joints in the granite. Minerals along the joints disintegrate into clays, more resistant quartz particles and rock fragments. Subsequent erosion removes the loose material leaving a pile of rounded boulders (after Trent and Hazlett, 2002).

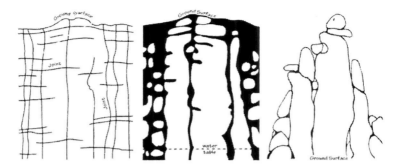

Figure 3. In other boulder piles, vertical joints may dominate and weathering produces more massive or vertically elongated bouldering (from the National Park Service, 2004).

In some areas, the granites are massive. A good example of this is at Barker Dam (Figure 4). A broad dome of granite is present on the eastern end of the impoundment, opposite from the dam. In massive granitic domes, which are originally formed deep below the surface, the removal of the

Figure 4. Granitic dome of the White Tank monzogranite at Barker Dam. This landscape is dominated by spalling, or exfoliation, of granitic sheets parallel to the dome surface (photo by D.D. Trent).

overburden releases stress causing curved joints generally parallel to the surface. When the pluton is exposed and weathers, great granitic sheets spall off the dome surface – similar to spheroidal weathering on a huge scale. The dome sheds, or exfoliates, these granitic sheets over time. This type of exfoliation dome is similar to that observed in Sierra Nevada landscapes such as Yosemite.

Granitic Sculpting by Ancient Soil Development

Ancient weathering related to soil formation and climate change has added another level of sculpting to the granites. One impact of ancient weathering is the presence of a notch, several feet above the current ground surface, which is visible on many granite outcrops. This notch is a remnant of weathering in a more humid environment. Previously, the Joshua Tree climate was much different from that of today. Temperatures were moderate with higher rainfall which promoted the formation of deep soil horizons. In the presence of more moisture, a thick soil column developed. Plants would grow and die adding organic material to the top soil layers. Moisture and organic acids collected in the upper horizon of the soil column. Where the soil impinged on the granitic masses, chemical weathering was accelerated along a band where the soil contacted the rock (Figure 5). As the climate changed, becoming hotter and drier, soil development ceased and eventually the thick accumulation of soil was eroded away exposing the notches. The removal of soils exposed the notches at heights up to several feet above the current ground surface. Such indentations are visible at Stop 4 along the 18-Mile Geology Tour, Hidden Valley picnic area and trail, Jumbo Rocks and elsewhere.

One additional weathering mechanism responsible for the Joshua Tree "look" is the presence of round, shallow indentations in many outcrops caused by cavernous weathering. These indentations are referred to as "tafoni," and may also be seen in sandstones and limestones, as well as granites elsewhere in the world. In Joshua Tree, they are responsible for formations like Skull Rock (Figure 6). Many of the tafoni in Joshua Tree appear to be aligned parallel to the outcrop-soil contact suggesting that their formation is related to former soil horizons.

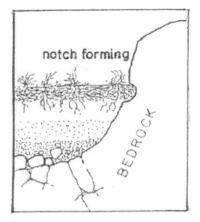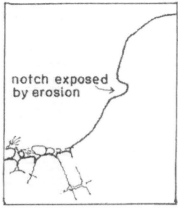

Figure 5. Development of notches observed in many granitic outcrops within Joshua Tree. During an earlier time of moist climate in the region, a thick soil profile developed. The notching occurs in a zone of intense chemical weathering just below the ground surface. As the climate changed and became drier, soil development ceased. The soil column that had existed previously was gradually weathered and eroded. The existing notch, which is now visible on many steep-sided rock surfaces as much as several feet above the present ground surface, reveals the position of the original soil surface (after Trent and Hazlett, 2002).

Figure 6. Skull Rock, visible along the south side of Park Boulevard in the Jumbo Rocks area is an example of cavernous weathering which produced round indentations in the rock surface called tafoni (photo by D.D. Trent).

In most cases, tafoni form due to the presence of moisture which leaches salts or minerals from within the rock, spalling off small bits of rock in the indentations where the moisture collects. While these types of forms are most common in coastal environments, it may be that during the day, deeper indentations in the granites offer some additional shade or protection for moisture which may condense after sunset. This cycle of condensation and evaporation may further deepen small indentations, resulting in patterns of rounded, concave depressions on the rock surface.

STRUCTURAL SETTING

Having discussed the factors which have sculpted the famous Joshua Tree granites, let's step back and put the Park in a broader geologic context. While the granites of Joshua Tree have inspired artists, challenged rock climbers and delighted geologists, there's much more here than just granite!

The Park lies at the easternmost end of California's Transverse Ranges. This mountain chain begins due west of Joshua Tree, where the Transverse Ranges meet the Pacific Ocean near Point Arbuello, 50 miles west of Santa Barbara. They end in the eastern reaches of Joshua Tree. Extending in an east-west direction, the Transverse Ranges are already unique for their orientation. Nearly all mountain ranges on our continent run north-south. But here, the North American and Pacific tectonic plates collide nearly head on, producing the east-west trend. Due to this plate interaction, the San Andreas fault system makes a dramatic, nearly right-angle bend along the southern edge of Joshua Tree and counties westward (Figure 7) to the "Big Bend." This tectonic setting, produced by north/south compression, has resulted in the east-west and northwest-southeast trending faults that both cut through and around the Park.

The left-lateral Pinto Mountain fault just falls within the Park boundary, beneath the Visitors Center. It then skirts along the western section of the northern park boundary (Figure 7). The total offset along the Pinto Mountain fault has been estimated to be up to 16 kilometers (Dibblee, 1968 and Hobson, 1998). Between Morongo Valley and Twentynine Palms, the Pinto Mountain fault generally follows State Highway 62 and is visible in side hill benches, spurs, and a possible stream offset.

In Twentynine Palms, the Pinto Mountain fault is expressed as a line of vegetation at the Oasis of Mara, just west of the Park Visitors Center. This is a classic

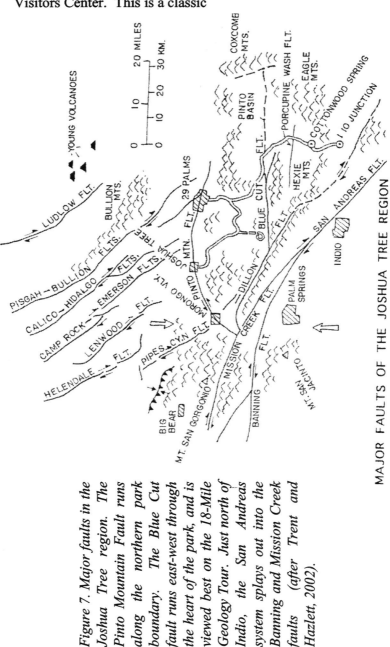

Figure 7. Major faults in the Joshua Tree region. The Pinto Mountain Fault runs along the northern park boundary. The Blue Cut fault runs east-west through the heart of the park, and is viewed best on the 18-Mile Geology Tour. Just north of Indio, the San Andreas system splays out into the Banning and Mission Creek faults (after Trent and Hazlett, 2002).

MAJOR FAULTS OF THE JOSHUA TREE REGION

example of a fault interrupting and controlling groundwater flow. Here, the subsurface flow is ponded on the fault's upslope side by the lower permeability of the buried fault surface, causing the water table to rise and create the oasis. A line of vegetation extends for about one and a half miles in conjunction with a three to six foot high fault scarp which is visible for about a half mile. Radiocarbon dating of detrital charcoal (Cardena et al., 2004) estimated the age of the most recent movement on the Pinto Mountain fault at 9400 years. Analysis of data suggests an active slip rate on the Pinto Mountain fault of 1.3 to 2.3 mm per year.

The Blue Cut fault is also left-lateral and runs generally parallel to the Pinto Mountain fault. The Blue Cut fault runs east-west, just south of Keys View, through Pleasant Valley and the Pinto Basin (Figure 7). The Blue Cut fault branches from the Dillon Fault zone, which runs northwest-southeast along the western boundary of the Park, adjacent to the Little San Bernardino Mountains. The Dillon fault essentially parallels the Mission Creek, Banning and San Andreas fault zone in this area.

Other minor faults occur throughout the Park but may not be dramatically visible. A minor fault appears to be responsible for the oasis at Cottonwood Springs in the southwestern portion of the Park at Cottonwood Campground. Similar to the Oasis of Mara, groundwater is apparently ponding up behind a vertical, lower permeability zone of fault gouge, which raises the water table to the ground surface. Other faults are also hinted at by observances of groundwater occurrence. Early in the 1960s, a groundwater survey of the Pinto Basin by the USGS noted that the groundwater elevation in two wells in close proximity to each other were at different depths (Kunkel, 1963). The difference is attributed to the presence of the Blue Cut fault lying between the two wells. This helped to place the hidden fault along the south side of the Pinto Basin.

ROCK TYPES EXPOSED IN JOSHUA TREE

Metamorphic Rocks

The oldest metamorphic rocks of Joshua Tree are believed to represent the remnants of a mountain range which existed on the early continental landmass known as Rodinia. Rodinia included present day North America, Australia,

Antarctica, Greenland and Scandinavia. Radiometric dating of these rocks suggests that they were metamorphosed well over one and a half billion years ago. As the landmass of Rodinia broke apart, fragments of the original mountain chain were widely dispersed. They survived in North America as portions of the Transverse Ranges, including the Pinto Gneiss of Joshua Tree. At least four subunits of the Pinto Gneiss are recognized: 1) the Joshua Tree augen gneiss is a granitic augen gneiss which outcrops in the Chuckwalla, central Eagle and south-central Pinto Mountains; 2) the metasedimentary suite of Placer Canyon is composed of quartzite and dolomite and unconformably overlies the Joshua Tree augen gneiss; 3) the distinctive augen gneiss of Monument Mountain, a dark colored prophyritic granodiorite-monzogranite which outcrops in the Hexie Mountains; and 4) the metasedimentary suite of Pinkham Canyon in the Chuckwalla, Eagle, Hexie and Pinto Mountains. The Pinkham Canyon suite includes quartzite, schist, fine-grained granofels and dolomite and is identical to a suite of rocks located in the northeastern-most Mojave Desert near Baker. This offers a suggestive link between the Proterozoic rocks of the Transverse Ranges and the North American craton. Radiometric dating yields ages of 1.65 to 1.7 billion years for the Joshua Tree augen gneiss and 1.65 to 1.68 billion years for the Monument Mountain augen gneiss (Barth, et al., 2004, this volume; Powell, 1982). Such dates make these some of the oldest rocks in California (Table 1).

Figure 8. Proterozoic gneiss exposed along the trail to the Lost Horse Mine (photo by D.D. Trent).

Geologic Age	Rock Units	Geologic Events
QUATERNARY (Holocene) *0.01 Ma*	dune sands, playa lake sediments, alluvial fans	10,000 years ago to present: faulting, uplift and earthquakes; weathering, mass wasting and erosion deposit sediment in valleys and canyons as present-day arid climate is established.
(Pleistocene) *1.6 Ma*	playa lake sediments, alluvial fans	Minor volcanism and major faulting in nearby Mojave Desert. Ice Age climate switches back and forth from glacial to temperate; alluvial fans form; pediments and inselbergs emerge as soil cover erodes.
TERTIARY *66 Ma*	alluvial fans deposited	*10Ma* to present: coastal ranges rise and Sierra Nevada creates rain shadow across Mojave region *10-25Ma*: warm, semi-arid climate produces savannah-like grasslands in Mojave region and thick soil horizon forms on bedrock in Joshua Tree region. *25-30Ma*: San Andreas fault zone forms as subduction ceases. *15-30Ma*: volcanism widespread throughout southern CA, possibly including lava flows in Hexie and Pinto Mts. *50Ma*: Erosion begins, widely exposing granitic plutons and gneiss; formation of plutons ends.
CRETACEOUS *144 Ma*	mafic & felsic dikes monzogranite of 49 Palms Oasis White Tank monzogranite (Figure 9) Queen Mt. monzogranite	Continued subduction and plutonism along the entire western edge of the North American plate. Note that Cretaceous granites make up the majority of the cores of today's Sierra Nevada, San Bernardino and Little San Bernardino Mts. (Age of Queen Mt. pluton is uncertain, may be Jurassic).
JURASSIC *200 Ma*	Gold Park Diorite	Continued subduction and plutonism.

Geologic Age	Rock Units	Geologic Events
TRIASSIC *251 Ma*	Twentynine Palms megacrystic quartz monzonite	*~250 Ma*: subduction of oceanic plate beneath North American plate initiates plutonism in western North America; Nevadan Mountains begin forming.
PALEOZOIC *545 Ma*	(no rock record)	Likely that marine sediments were deposited, however these would have been removed by subsequent erosion.
PRECAMBRIAN/ **PROTEROZOIC**	Gneisses, marble, quartzite	*800 Ma*: breakup of Rodinia, early continental landmass that included present day North America, Australia, Antarctica, Greenland and Scandinavia. Joshua Tree area likely continental shelf receiving terrigenous sediment from early Rodinian mountain ranges. *1871-1650 Ma*: Rodinian mountains form in association with metamorphism and plutonism.

Ma = one million years

Table 1. Summary of the major geologic events and associated rock units in the Joshua Tree region (after Trent and Hazlett, 2002).

Figure 9. Inselbergs (erosional remnants) of White Tank monzogranite at Hidden Valley (photo by D.D. Trent).

Igneous Intrusives

A minimum of five major plutons have intruded the metamorphic rocks described above. These plutons range in age from the middle Proterozoic to the Cretaceous (Table 1). The oldest are a series of igneous protoliths that intruded the metasedimentary rocks or their unmetamorphosed precursors. These intrusive rocks include foliated hornblende meta-gabbro; diorite and amphibolite; laminated orthogneiss and various leucocratic granite orthogneiss which are commonly garnetiferous. Magmatic episodes occurred at about 1780, 1750 and 1690-1675 Ma (Barth, et al., 2004, this volume). A younger Proterozoic suite of plutonic rocks intrudes the older Proterozoic gneisses in the Eagle Mountains (Powell, 1982) and are correlated with the San Gabriel anthorosite-syenite-gabbro complex, dated at 1191 ± 4 Ma (Barth, et al., 1995, 2000; and Powell, 1982).

The Triassic and Cretaceous plutons in the park, similar to granites of other California terranes including the Sierra Nevada, Klamaths and White-Inyos, are thought to have been generated during oceanic-continental plate convergence. Intrusive contacts between these plutons with the Proterozoic gneiss are exposed along the trail to the Forty-nine Palms Oasis north of Queen Mountain, and along the eastern edge of the Lost Horse Valley on the west slope of Ryan Mountain.

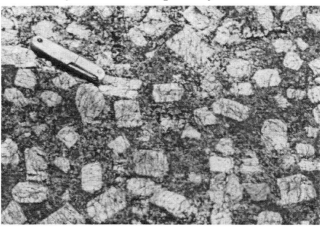

Figure 10. Potassium feldspar phenocrysts, up to two-inches in length, occur in the Twentynine Palms quartz monzonite (photo by D.D. Trent).

The Twentynine Palms megacrystic quartz monzonite is the oldest of the Mesozoic intrusions. This particular rock has a fine-grained matrix which surrounds large potassium feldspar phenocrysts that can be up to two inches in length (Figure 10). Preliminary age dating of this unit indicates a Triassic age of 240-245 Ma (J.L. Wooden, pers. comm., 2001). This megacryst pluton is one of a widespread group of Permo-Triassic plutons that are exposed in Southern California. These plutons are important because their presence indicates the convergence of oceanic and continental plates, and the subsequent generation of these plutonic bodies along the resulting subduction zone. The megacrystic Twentynine Palms quartz monzonite can be found along the trail to the Fortynine Palms Oasis, and along the arroyo east of Indian Cove Campground.

Cretaceous-aged plutonic bodies include the Queen Mountain monzogranite, Gold Park diorite, White Tank monzogranite, and the Oasis Monzogranite (Table 1). These plutons may be related to intrusive events in the Sierra Nevada, Mojave Desert and Peninsular Ranges. The oldest of these intrusive bodies is the Queen Mountain pluton. This monzogranite is comprised of coarse-grained quartz and potassium feldspar plus biotite or hornblende. The age of the Queen Mountain pluton is uncertain, it may be as old as Jurassic (J.L. Wooden, personal communication).

In many of the more accessible areas of the Park, one of the most visible rocks is the White Tank monzogranite. The White Tank is similar to the Queen Mountain monzogranite, but is finer-grained and contains biotite and/or muscovite but with no hornblende. Some of the most picturesque and well-known areas of the Park including Indian Cove, Jumbo Rocks, the Wonderland of Rocks, and Lost Horse Valley are outcrops of the White Tank (Figures 1 and 9).

The Oasis of Fortynine Palms monzogranite is the youngest of the Cretaceous plutons. This garnetiferous, muscovite-bearing pluton is exposed near Fortynine Palms Oasis. The garnets included in this pluton are small and dark red. The muscovite is present as small crystals within the rock mass.

Some smaller dark plutons also occur, collectively named the Gold Park diorite. These small plutons yield a Uranium-Lead (U-Pb) age of about 165 Ma (J.L. Wooden, personal communication, 1999). Younger dikes of varying compositions including felsite, aplite, pegmatite, andesite and diorite cut across all of these plutonic bodies. Although the pegmatites are compositionally similar to the granitic plutons, the large crystal size makes them distinctive.

Some of the youngest geologic materials have had perhaps the most influence on the history of the Park. Milky quartz veins can be found throughout the Park, and their association with metallic minerals – most notably gold – have been responsible for the significant mining operations that are an important part of the Park history. These quartz veins are often also associated with sulfides such as pyrite. Chemical weathering of pyrites commonly stains the surrounding rock materials and was often a clue to hopeful miners that other more important metallic minerals might be present in the quartz veins.

Joshua Tree Basalts

Although limited in extent within the park, basalts form some of the more interesting features on the landscape, and have been the subject of much ongoing study.[1] There are at least ten basalt centers in the Park which appear to be associated with major east/west fault traces including the Blue Cut and Eagle Mountain faults (Barth, et al., 2004, Figure 1; Probst, et al., 2004, this volume). Probably the best known basalt outcrop is that of Malapai Hill, located on the southern edge of Queen Valley, near Stop 7 of the 18-Mile Geology Tour.[2] Malapai Hill is a basalt plug that has been revealed by erosion. Mantle-derived inclusions of lherzolite are found both in the Malapai Hill basalt and in the basalt outcrop in Lost Horse Mountain. The presence of the olivine-rich peridotite suggests that these basalt magmas have risen 30 to 50 miles through the earth. Both the Lost Horse and Malapai Hill basalts are composed of massive, tilted columnar basalt. Basalt outcrops on the sides of Malapai Hill and Lost Horse Mountain show strong columnar jointing (Figure 11).

Other basalts in the more southern and eastern portions of the Park area associated with the Victory Pass and Porcupine Wash faults (Probst, et al., 2004, Figure 1, this volume). The more eastern basalts tend to be composed of multiple flows, often with dense cores and vesicular texture in the uppermost flows. These eastern basalts have not been found to contain lherzolite inclusions.

---

[1] Probst, et al. (2004, this volume) have reported on the ongoing studies of the timing and geochemistry of the Joshua Tree basalts.

[2] See Road log for 18-Mile Geology Tour, this field guide

Figure 11. Basalts of Lost Horse Mountain showing strong columnar jointing and tilted orientation (photo by D.D. Trent).

The Patina of an Arid Climate

Most rock outcrops within the Park exhibit some degree of rock varnish. These coatings of manganese and iron oxides have darkened many rock surfaces but they have also provided a canvas for early human occupants to leave their marks in petroglyphs scattered throughout the area. The darker, brownish-black coatings are enriched in manganese. The more reddish-brown colors originate from the weathering of iron oxides. Rock varnish can originate from the concentration of iron and manganese by microbial action on the rock surface, or by the deposition of iron and manganese oxides with atmospheric clays from fine dust particles (Dorn and Oberlander, 1981). A study of petroglyphs in the adjacent Salton Sea area (Turner and Reynolds, 1977) determined these carvings were etched into the rock varnish some 9,100 years ago.

SUMMARY

Within its boundaries, Joshua Tree National Park holds a wealth of rock types and landforms. Both the seasoned geologist and the casual naturalist will find much to inform, educate and inspire. Over the past decades, the landscape of Joshua Tree has provided an outdoor classroom for students of geology, botany, biology, archeology and the environment. In our summary here, we have only skimmed the surface of all this desert gem has to offer. We hope it will encourage you to read more and do your own exploring!

REFERENCES AND ADDITIONAL READING

Barth, Andrew P., Wooden, Joseph L., and Jarvis, Janet L., 2004. *Crust Formation And Evolution In Southern California: Field And Geochronologic Perspectives From Joshua Tree National Park, in*: Mining History and Geology of Joshua Tree National Park, San Diego Association of Geologists Annual Field Guide, 2004., Margaret R. Eggers, editor. Sunbelt Publishing, San Diego California.

Barth, Andrew P., Wooden, J.L., Jacobson, C.E., and Probst, K., 2004b, *U-Pb geochronology and geochemistry of the McCoy Mountains Formation, southeastern California: A Cretaceous retroarc foreland basin,* Geological Society of America Bulletin 116, 142-153.

Barth, Andrew P., Wooden, Joseph L., Coleman, D.S., and Fanning, C.M., 2000. *Geochronology of the Proterozoic basement of southwestern most North American, and the origin and evolution of the Mojave crustal province.* Tectonics, vol. 19, 616-629.

Barth, Andrew P., Wooden, Joseph L., Tosdale, R.M., Morrison, J., Swason D.L., and Hornley, B.M., 1995. *Origin of gneisses in the aureolo of the San Gabriel anorthosite complex and implications for the Proterozoic crustal evolution of Southern California.* Tectonics, vol. 14, 736-752.

Cadena, Ana M., Rubin, Charles M., Rockwell, Thomas K., Walls, Christian, Lindwall, Scott, Madden, Chris, Khatib, Faten, and Owen, Lewis, 2004. *Late Quaternary Activity of the Pinto Mountain Fault at the Oasis of Mara: Implications for the Eastern California Shear Zone.* GSA Abstracts with Programs; 2004 Annual Meeting, Denver, Colorado.

Dibblee, T. W., Jr., 1968. *Geologic Map of the Twentynine Palms Quadrangle, San Bernardino and Riverside Counties, California,* US Geologic Survey Investigations Map I-516.

Dorn, R.I., and Oberlander, T.M., 1981. *Microbial origin of desert varnish,* Science, vol. 213, no. 11, 1245-1247.

Kunkel, Fred, 1963. *Hydrogeologic and Geologic Reconnaissance of Pinto Basin: Joshua Tree National Monument, Riverside, California.* US Geologic Survey Water-Supply Paper 1475-O.

Oberlander, T.M., 1972. *Morphogenesis of granite boulder slopes in the Mojave Desert, California.* Journal of Geology, vol. 80, no. 12, 1-20.

Powell, Robert E., 1982. *Crystalline basement terranes in the southeastern Transverse Ranges, California. in:* Cooper, J.D., editor, Geologic Excursions in the Transverse Ranges: Geological Society of America Cordilleran Section, 78th annual meeting, Anaheim, California, 109-136.

Probst, Kelly P., Barth, Andrew P., and Yi, Keewook, 2004. *Neogene volcanism in Joshua Tree National Park, in*: Mining History and Geology of Joshua Tree National Park, San Diego Association of Geologists Annual Field Guide, 2004, Margaret R. Eggers, editor. Sunbelt Publishing, San Diego, California.

Trent, D.D., 1998. *Geology of Joshua Tree National Park,* California Geology, v. 51, no. 5, October.

Trent, D.D., and Hazlett, Richard W., 2002. *Joshua Tree National Park Geology,* published by the Joshua Tree National Park Association.

Turner, W.G., and Reynolds, Robert, 1977. *Dating the Salton Sea Petroglyphs,* Science News vol. 213, no. 11 (September), 1245-1247.

United States Dept. of the Interior, National Park Service, 2004. *Joshua Tree Guide: A Planning Guide for Visitors to Joshua Tree National Park,* Spring, 2004.

Wooden, J.L., Fleck, R.J., Matti, J.C., Powell, R.E., and Barth, A.P., 2001, *Late Cretaceous intrusive migmatites of the Little San Bernardino Mountains, California*, Geological Society of America, Abstracts with Programs 33, no. 3, A65.

INSIDE FRONT COVER: *Geology of Central Part of Joshua Tree National Park*. Map prepared by D.D. Trent.

Joshua Tree Shadow II (Photo by John D. Clark, 2004).

GEOLOGY AND HISTORY OF MINES OF JOSHUA TREE NATIONAL PARK

D.D. "Dee" Trent, PhD, *Professor Emeritus,*
Citrus College, Glendora, CA 91741

INTRODUCTION

Mining is an integral part of the history of the Joshua Tree National Park region. There are 288 abandoned mining sites in Joshua Tree National Park with 747 mine openings.[3] Gold was the commodity of greatest interest. In the Pinto Mountains, immediately outside the Park, are two mining districts with numerous abandoned mines. As recently as 1998, there were eight claimants having mining claims in the Park. The sites include mill and mine sites, gravel pits, some open pits, but most were underground operations (Chris Holbeck, personal communication, 1998).

Mining activity in the region began in the 1870s, but peaked in the 1920s and 1930s. The ore produced by the many mines came mainly from gold-bearing quartz veins that intruded Mesozoic granitic rocks and Proterozoic metamorphic rocks (Tucker and Sampson, 1945; Trent, 1998). Figure 1 shows the various mining districts within the Park.

---

[3] A mine site may include several openings with some combination of shafts, open cast cuts and adits. Some sites included mills. The most recent records available show that in 1998, two claimants still held eleven patented claims, meaning the claimants hold titles to the property. These patented claims are within the old (former) Monument boundary. Six claimants held 49 unpatented claims, all in the new lands added through the California Desert Protection Act (CDPA).

Park lands are withdrawn from mineral entry, which excludes claim staking. Existing claims, however, are grandfathered in and claimants may conduct mining activities at their claims. They must first comply with the Mining in the Parks Act (1976), and requirements of Code 36 of Federal Regulations, Part 9A. These regulations ensure that an environmentally considerate operation takes place on a legal claim, where reclamation design is a requirement of the operation.

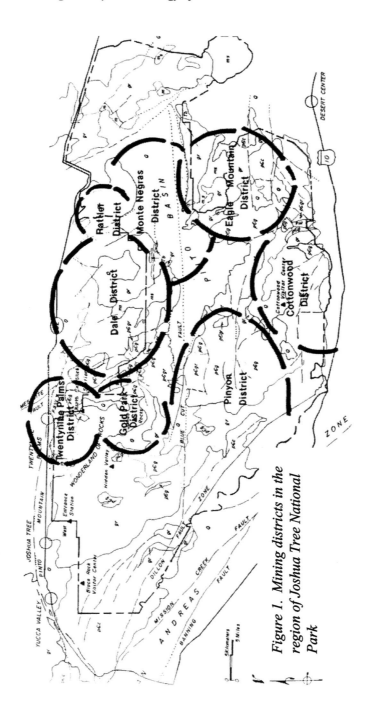

Figure 1. Mining districts in the region of Joshua Tree National Park

There are eight mining districts in and adjacent to Joshua Tree National Park, the: Twentynine Palms, Dale, Rattler, Monte Negras, Eagle Mountains, Cottonwood Spring, Pinyon, and Gold Park. The mines most easily visited are the Mastodon in the Cottonwood Spring District, the Lost Horse, Gold Coin and the Silver Bell in the Middle Pinyon District, and the Desert Queen in the northern Pinyon District.

DESERT QUEEN MINE (McHaney Mine)

Location

The Desert Queen Mine (Figure 2) is on the northeastern margin of Queen Valley. It is reached by a well-worn, three-quarter-mile trail that begins at the Pine City Back Country parking area.

Years of operation

1895 - 1942? (Greene, 1983, p. 204)

Production:

1895-1900 (?): 3,701 oz. gold (worth ~$1,454,500 in July, 2004 dollars).

1900-1941: 144 oz. gold, (worth ~$56,600 in July, 2004 dollars) and 87 oz. silver (worth ~$523 in July, 2004 dollars).

It is reported by Willis Keys that his father, William F. "Bill" Keys, made more money by leasing out the mine to operators than he ever earned from his own operation. The terms of a lease would be for one year but the operators usually walked away from the mine before the lease expired (Bill Truesdell, personal communication, 2002).

Geology

The Desert Queen Mine is developed in the White Tank monzogranite, south of the contact with the Palms Quartz monzonite. Gold is described as being in pockets associated with dikes of pegmatite, aplite, andesite, and veins of gold-bearing quartz. Minor veinlets of malachite and azurite, pyrolusite dendrites, and chrysocola occur in fractures in some of the mineralized pockets in the aplite and quartz veins. Waste rock piles are stained with these secondary minerals. The ore was milled at the McHaney Ranch (which eventually was owned by Bill Keys and became the Desert Queen Ranch).

Figure 2. Desert Queen Mine, adit portal and cyanide tanks circa 1935 (Photo from the collection of the National Park Service).

History

The story of the mine's discovery in 1895 reads like a wild west dime-novel involving a hapless prospector, a gang of thieves, a cold-blooded murder and a trial that is a classic example of poor justice. The story is told in Bob Cates' book, *Joshua Tree National Park: A Visitor's Guide,* and in Keys and Kidwell's *Growing Up At the Desert Queen Ranch.* Keys and Kidwell relate that the jury's verdict was that James "came to his death from two pistol shot wounds inflicted by Charles Martin ..." and that Martin had "acted in self-defense while his life was in jeopardy at the hands of the deceased," who supposedly was attacking Martin with a knife. Inexplicably, a pistol just happened to be lying on the ground near Martin which he picked up and shot James (Keys and Kidwell, 1997, p. 67-69).

In the 1960s, when Bill Keys was being interviewed about the region's early history, he related a very different version of the acquisition of the Desert Queen Mine by the McHaney Gang. "James was a miner working in the Lost Horse Mine, but on Sundays he'd walk and prospect. Well, these coyote cowboys (George Meyers, Charlie Martin, and James

McHaney) saw his track because it was easy to follow. They were as sharp as a coyote as to any disturbance in the ground."

"So these three fellows came here on horseback and stayed up on a hill above James' cabin off the Lost Horse Mine road. Charlie Martin walked down a little ways and yelled to James who was in his cabin in the ravine. 'Do you own this? We found a rich strike here.' And he (James) went up to look, and Martin shot him. That was to get the Queen Mine. That was up on the ridge above James' cabin that they shot him" (Keys and Kidwell, 1997,p. 67-69). [There's an obvious problem with this story: the Lost Horse Mine road is about 10 miles from the location of the Desert Queen strike. Was James' cabin in the ravine below the Desert Queen Mine or in the ravine along the Lost Horse Mine road? – *DDT*].

Keys reported that Jim McHaney paid Martin $47,000 from the first mill run of Desert Queen ore for his involvement, and he gave George Meyers a herd of cattle for his involvement in the shooting. James' remains were buried in the little cemetery on the Ryan Ranch near today's Ryan Campground. The mine went through a series of ownership changes and eventually became the property of Bill Keys.

The McHaney Gang

In the 1870s and 1880s, James B. and William S. McHaney ran cattle in upper Santa Ana Canyon near Seven Oaks, in the San Bernardino Mountains. Their herds were noted to grow in size mysteriously and it became obvious that they were rustling cattle. Their brand was never recorded, as required by law, there being so many variations of their brand from crude rebranding of rustled cattle, a single registered brand was impossible.

Jim McHaney, the more ruthless law breaking brother, collected four other like-minded cowboys, George Myers, Charlie Martin, Willie Burton, and Ike Chestnut, and they became known as the McHaney Gang. Growing tired of the gang's rustling cattle from Bear Valley, Redlands and Highland, a posse of East San Bernardino Valley ranchers was organized and tried to capture them. Although unsuccessful, they did manage to chase them out of the San Bernardino Mountains.

The McHaney Gang...continued

Thus, in 1888, the McHaney Gang moved their cattle business to the Little San Bernardino Mountains near modern day Twentynine Palms. They continued cattle rustling, hiding their cattle in a box canyon at their headquarters, Cow Camp, on the western edge of the Wonderland of Rocks. It was here that McHaney turned his eye toward gold mining (Robinson, 1989, p.89-91).

The gang discovered that Frank L. James, who worked for the Lang family at the Lost Horse Mine and prospected on his days off, had discovered the Desert Queen lode in 1894. At this point the story becomes confused, there being at least three different uncertain scenarios explaining the suspicious circumstances by which Jim McHaney gained control of the Desert Queen by claim jumping, with Frank James being killed at the hand of Charlie Martin. Bill McHaney, the more law abiding brother was not involved in the takeover of the Desert Queen, lived a respectful life spending part of his time at the Oasis of Mara (Twentynine Palms) and at a cabin in Music Valley (in the Pinto Mountains) where he was prospecting as late as the 1920s (Daily Sun, 1960). Jim McHaney eventually turned to counterfeiting which landed him a seven-year term in a federal penitentiary. Following that it is reported that he worked for the Los Angeles street cleaning department (Cates, 1995, p. 75; Daily Sun, 1960).

References

Cates, Robert B., 1995. *Joshua Tree National Park: A Visitor's Guide*; Chatsworth, CA, Live Oak Press.

Daily Sun (San Bernardino), 1960. *Joshua tours lead over historic area*; August 8 issue.

Robinson, John W., 1989. *The San Bernardinos*; Arcadia, CA, Big Santa Anita Historical Society.

GOLD COIN MINE (Gold Galena Mine)

Location

The Gold Coin Mine (Figure 3) is on the south flank of the Hexie Mountains along the Geology Tour Road in Pleasant Valley, adjacent to the Geology Tour Backcountry Board, about seven miles south of Park Boulevard.

Years of operation

1900?-1933 (Greene, 1983, p. 252-253).

Production

Unknown

Geology

The ore is in stringers in an east-west shear zone containing thin, discontinuous quartz in an irregular body of granite within the Precambrian Augustine Gneiss (Ruff, et al, 1982, p. 229). The Augustine gneiss consists of retrograded granulites ranging in composition from tonalite to granite (Powell, 1980). With only one exception, the Hexahedron Mine, four miles north east of the Gold Coin, all mines in the Middle Pinyon, or Hexie, district follow oxidized veins of milky quartz occurring along faults within the gneisses. Most faults tend to follow the regional northwest to west trend, and are likely related to the Blue Cut fault.

The significance of the Blue Cut fault and other faults in the region in their relation to mineral occurrences may lie in the fact that the extensive fracturing and brecciation associated with these faults provided a plumbing system for oxygenated meteoritic water to percolate into low-grade sulfide metal deposits which could be leached to enrich the deposit and form an economic ore.

History

Information on this mine is scanty. The site was discovered in 1900 and developed by the German-American Mining and Milling Co. in the early 1900s. The directors were F.C. Longnecker, S.L. Kistler, and A.N. Hamilton. By the fall of 1908 it reportedly was working three mines, the Texas Chief, Lone Star, and the Apex. In 1911 it reported that water was hauled to the mines from Pinyon Well in Pushwalla Canyon (Greene, 1983, p. 252-253).

Figure 3. Remains of the mill at the Gold Coin Mine (Photo by Robert Cates).

Events on record (from Greene, 1983, p. 252-253):

1916, the mine was bought out by the Gold Galena Mining Co. with underground workings from 70 to 100 ft. on a vein containing galena-carrying gold. No mill is mentioned. The mine closed in 1918.

1920, Bill Keyes located two claims in the area, but they were unexplored. No real development.

1922, six claims and an inclined shaft 100 ft. deep. No record of production.

1929, a mill of sorts was located here with five claims owned by Longnecker with two men employed; a shaft 50 ft. deep is reported. Gold values averaged $12 per ton (Greene, 1983, p. 253). Probably part of the 1930s effort was in processing old tailings which was being done in many old mining camps in the west during the years of the Great Depression. The tank that still remains at the mill site was for storage of either water or cyanide.

1957, Cliff Gray (personal communication, 1998), geologist with the California Division of Mines and Geology, reported eight inclined shafts and two vertical shafts in the area.

SILVER BELL MINE

Location

The Silver Bell Mine (Figure 4) is on the east end of the Hexie Mountains. It is the most obvious mine in the park, with two ore bins clearly visible from milepost 8 on the Pinto Basin Road. The site is accessible by walking cross country a short distance until joining the obscure abandoned mine road. Following the road-trail about 0.5 mile brings one to the site. Remains of buildings and various artifacts exist along the road-trail.

Years of Operation

1934 – 1962 (Emerson, 2000)

Production

Known production 1934-1954, 219 oz. gold, and 53 oz. silver (Emerson, 2000).

Geology

The workings follow a N60W trending 4-ft wide fault zone and several minor faults containing oxidized gold-bearing quartz veins within Augustine Gneiss. Assays in 1958 revealed low values of silver and gold but copper values at ~$90 per ton (Ruff, et al, 1982, p. 229; Emerson, 2000).

History

Little is known about the history of the mine. In the 1930s it was a gold mine. Upon closure of gold mines by the Federal government during World War II in order to release miners for work directed at the war effort, the Silver Bell became a lead mine. From 1956 to 1962 it was operated by the Farrington-Mann Company as a copper property (Emerson, 2000).

Figure 4. Ore bins at the Silver Bell Mine, Hexie Mountains (Photo by D. D. Trent, 2001).

LOST HORSE MINE

Location

The Lost Horse Mine (Figures 5 and 6) is near the summit of Lost Horse Mountain, between the southern end of Hidden Valley and Queen Valley. Starting from the Lost Horse Mine Parking area, a two-mile trail follows the old mine access road to the mine.

Years of Operation

1894 – 1942? (Ruff and others, 1982, p. 231)

Production

Minimum production estimates are 10,500 oz. of gold and 16,000 oz. of silver (Fife and Fife, 1982, p. 455-456). In 2004 dollars those yields amount to about $4,127,000 in gold, and $96,300 in silver.

Geology

Sampson and Tucker (1945, p. 137) report the ore body is one or more quartz veins that occur in micaceous quartzite and granite. However, the host rock is clearly granitic quartz biotite gneiss; the Proterozoic Lost Horse pelitic granofels of the

Pinto Gneiss according to Powell (1980). Coleman, et al, (2003, p. 67) map it as orthogneiss and with U-Pb ages of ~1715 and ~1700 Ma. The gold-bearing quartz vein in the mine varies in width from six inches to five feet (Sampson and Tucker, 1945, p. 137).

Figure 5. Lost Horse Mine, 1977. Photo shows ten-stamp battery, Holt gasoline engine and ore bin (Photo by D.D. Trent).

History

As with the Desert Queen Mine, the details of the discovery of the Lost Horse Mine is a bit cloudy. One story is that the mine was discovered in 1893 by four men: George Lang, John Lang, Ed Holland and James Fife. The other story, perhaps more likely, is that young Johnny Lang, with his father and brother, brought a herd of cattle from Texas to Southern California in the early 1890s. On arriving in the Lost Horse area Lang met "Dutch" Diebold, a prospector, who had discovered likely-looking gold-bearing quartz near Pinyon Mountain but he had been driven away from the site by the McHaney Gang. Young Lang, too, experienced a similar encounter with the Gang. While looking for a lost horse, Lang walked into McHaney's camp where he had a run-in with Jim

McHaney (Cates, 1995, p. 62). Lang offered Dutch $1,000 for his claim if it proved out. Lang began developing the property but under the ever-present threat of being killed by the McHaney outlaws. Meanwhile, young Lang's father had taken up residence at Witch Springs (now Lost Horse Well) and advised his son to take in some partners. Despite variations in the story, it is verified by official county records that George Lang, John Lang, James Fife and Ed Holland filed a location notice for the property in December, 1893 (Greene, 1983, p. 256).

Development began in 1894 when rich ore was hand cobbed from ore-shoots in the Lost Horse vein. Rich outcroppings and float of this gold were also found, some of this rich ore being sold as specimen gold (Fife and Fife, 1982, p. 458). The richest known specimen of gold found near the mine was picked up by Jim Fife. It was a mass of gold the size of a man's fist; the grade of the ore estimated to run 4,000 oz per ton. Pieces of this gold-quartz nugget are still in the Fife family (Fife and Fife, 1982, p. 458). In the early stages of development, the high grade milling ore was processed by a two-stamp mill at Pinyon Well. This soon proved to be unsuitable causing Lang and Fife to erect their own two-stamp mill at Lost Horse Spring. In 1897 the mine was patented, a new ten-stamp mill was installed at the mine and a five-mile pipeline built to the site from Lost Horse Spring (Greene, 1983, p. 256).

Sampson and Tucker (1945, p. 137) report the development of the mine includes an 80-foot tunnel driven on the Lost Horse vein, and a 500-foot shaft sunk on the vein with drifts at the 100, 200, 300, and 400 levels.

Gasoline power was substituted for steam in the 1920s, and the last work appears to have been done in 1936 when the Ryans, or a Mr. Phelps, treated approximately 600 tons of tailings with cyanide. Also, in 1936, pillars of ore were removed from the upper levels and milled at the property by the 10-stamp mill. Despite all of the work done in the 1930s, only a few hundred ounces of gold were recovered during the decade (Sampson and Tucker, 1945, p. 137; Greene, 1983, p. 256-257).

The Lost Horse Mine was acquired by the National Park Service (NPS) from the Ryan descendants in 1966. The mine road was closed to vehicles, the mill restored as a prime interpretative exhibit, the head frame taken down, and the mine shaft sealed by a concrete slab (Green, 1983, p. 261).

Figure 6. Lost Horse Mine, primary crusher, headframe and ore bin, circa 1958 (Photo by Bruce W. Black, collection of the National Park Service).

MASTODON MINE

Location

The Mastodon Mine is located in the Cottonwood Mountains and can be reached easily by hiking a three mile loop trail from the parking lot at Cottonwood Spring. The trail passes the mine and the remains of the Winona Mill.

Years of Operation

1934 - 1971 (Greene, 1983, p.170).

Production

Unknown

Geology

Tucker and Sampson (1945, p. 138) report the mine followed three parallel quartz veins in granite. The vein widths ranged from eight to twelve inches.

History and Development

Original location recorded in November, 1934. Development consists of an inclined shaft to a depth of 45 feet. Milled ore is reported to have averaged $40 per ton in gold. The mine was owned by the Hulsey family of Indio who performed the required assessment until 1971 when the property was acquired by the NPS (Greene, 1983, p. 170; Tucker and Sampson, 1945, p. 138).

WINONA MILL

(aka Cottonwood Springs Custom Mill)

Location

The Winona Mill site is located on the west end of Cottonwood Mountains and is accessible by following the Mastodon Trail from Loop A of the Cottonwood Campground for about 0.5 mile; this is a section of the Mastodon Mine loop trail.

Years of Operation

Initial date of operation unknown but probably the mid-1930s – 1945 (?) (Greene, 1983, p. 170).

History

The mill was operated by the Hulsey family of Indio in conjunction with the development of their Mastodon Mine. The concrete foundations that remain supported the mill's machinery. Ore was processed with a Fulton jaw crusher, drag classifier, amalgamation plates and an amalgamation trap. Pulp from the trap was pumped to a cone classifier with overflow fed to a ball mill. Slurry from the bottom of the cone was routed either to a concentrator or a 2-cell Groch flotation machine; tailings from the concentrator were routed to a settling tank, and the water from the settling tank returned to the ball mill. A 40 horsepower Buick engine powered the mill. The mill had a capacity of 40 tons per day with a recovery of 85 percent of the gold values (Sampson and Tucker (1945, p. 129).

The location for the Winona Mill was selected because the site is one of only two springs between Mecca and Dale; water, of course, being critical in milling ore. The mill processed contract ore from smaller mines in the Hexie Mountains as well as ore from the Hulsey's Mastodon Mine (Moore, date unknown). In addition to the mill, this was the residence for workers who planted non-native shrubs and trees and cottonwoods which still flourish at the site (Furbush, 1995, p. 123).

WALL STREET MILL

Figure 7. Two-stamp mill at Pinyon Well, circa 1920. Bill Keys later moved the stamp mill to the Wall Street Mill (Photo from the collection of the National Park Service).

Location

The Wall Street Mill (Figures 7 and 8) is at the entrance to Wall Street Canyon near the southeastern edge of the Wonderland of Rocks. It is reached from Wonderland Ranch Parking Area by following the well-worn trail that is the abandoned Wall Street Mill road about one mile to the site.

Years of Operation

~1928 – 1949 (Greene, 1983, p. 218)

History

The site began as a 31-foot-deep well supposedly dug by Bill McHaney in 1896 for watering his livestock. Sometime later George Meyers took control of the well and also used it for cattle. The well was deepened to 50 feet. In 1928, Oran Booth and Earl McGinnis needed a mill to work ore from their Wall Street Mine west of Jumbo Rocks and, because of the need for water to process the ore, they located their mill adjacent to Meyer's Well. The mill was given the same name as their mine. But Booth and McGinnis' dissolved their partnership when their mine played out and they offered the mill to Bill Keys who filed a milling claim on the site in 1930.

A family in the area named Oberer arranged with Bill Keys to purchase the Wall Street Mill site, a mine, and the mill at Pinyon Well. They moved the two-stamp mill from Pinyon Well (located in Pushwalla Canyon southwest of Pleasant Valley) to the Wall Street Mill site.[4]

The Oberer's mining efforts proved unsuccessful and they left with Bill Keys again taking possession of the mill. With the help of a miner named Hopper, who had need of a mill to process ore from his mine, Keys rebuilt the mill in 1932 so it could be operated by one man (Keys and Kidwell, 1997, p. 72).

The site included two other structures in addition to the mill: a bunkhouse, now gone, and an outhouse, now collapsed. The mill building is framed from heavy timber salvaged from other mine structures in the region, and the framing was done

---

[4] The mill at Pinyon Well was probably the first mill to be erected in the region. The presence of a reliable, year-round supply of water necessary for operation dictated this original site. Also, it was readily accessible because it was situated on the Pushwalla Canyon freight road, the only route in the late 19th century that served the region that is now Joshua Tree National Park. Thus, it was convenient for miners in the region to bring their ore to the mill for custom milling. The mill was operated by E. Holland & Company, and the two-stamp mill was manufactured in Los Angeles by the Baker Iron Works in 1891.

largely by a woman millwright, a Mrs. Hopper. The structure is covered with corrugated iron. A 12-horsepower Western gas engine powered the belt-driven Fulton jaw crusher, Baker Iron Works two-stamp mill, and Myer concentrating table by a system of line shafts and pulleys that is typical of 19th century technology (Greene, 1983, p. 240). The structure and the machinery are still in good condition, although the copper plates are missing from the amalgamation table.

Bill Keys operated the mill doing custom work for other mines in the area more-or-less regularly during the 1930s and then episodically until 1943. At times the mill ran for 24 hours a day with a maximum daily output of two to five tons depending on the nature of the ore. He charged miners $5 a ton for custom processing their ore which ran from $35 to $50 a ton (Keys and Kidwell, 1997, p. 72-75). Keys' operation of the mill ended in 1943 when he was convicted of manslaughter in the killing of Worth Bagley. Consequently, Keys, at age 64, was sent to San Quentin. He was paroled in 1948 and received a full pardon in 1956 as a result of an investigation and magazine articles written by Earle Stanley Gardner (author of the Perry Mason mystery series) about Keys and the unjust conviction (Cates, 1995, p. 56-57).

Willis Keys operated the mill from 1947 until 1949 when it was shut down. Willis stated that at some time, probably in the late 1930s, someone cyanided the mill tailings. The mill was operated again, but only briefly, in 1966 (Greene, 1983, p. 218).

Bill Keys died in 1969 and the site was relinquished to the NPS by the Keys' estate in 1971. In 1975, the Wall Street Mill was added to the National Register of Historic Sites due to its technological significance (Greene, 1983, p. 218). It is a fine example of a small but complete gold ore amalgamation mill featuring 19th century technology, one of the few in the West that still stands where it was used.

Figure 8. Exterior of the Wall Street Mill. Ore was delivered into the grizzly and crusher by ore cart along the inclined tracks on the right. The metal framing extending above the roof is the upper support and cam assembly for the two stamps (photo by D.D. Trent).

REFERENCES AND ADDITIONAL READING

Cates, Robert B., 1995. *Joshua Tree National Park: A Visitor's Guide*; Chatsworth, CA, Live Oak Press.

Coleman, D.S., Barth, A.P., and Wooden, J.L., 2003. *Metamorphism of orogneiss of Proterozoic rocks of the southwestern Mojave Province*: Geological Society of America Abstracts with programs, 35, no. 4, p. 67.

Emerson, John W., 2000. *Mines in Joshua Tree National Park*, 1998-1999 Report, Masters thesis in library at Joshua Tree National Park.

Fife, E. J., and Fife, D.L., 1982. *Geology and mineral resources of the Lost Horse gold mine, Lost Horse quadrangle*, Riverside County, California, *in* Fife, D.L., and Minch, John A., editors, Geology and mineral resources of the California Transverse Ranges: Santa Ana,. South Coast Geological Society, p. 455-465.

Furbush, Patty A., 1995. *On foot in Joshua Tree National Park*: Lebanon, Maine, M.I. Adventure Publications.

Greene, Linda W., 1983. *Historic resource study: a history of land use in Joshua Tree National Monument*; Denver Service Center, National Park Service manuscript.

Keys, Willis, and Kidwell, Art, 1997. *Growing up at the Desert Queen Ranch*; Twentynine Palms, Joshua Tree National Park Association.

Moore, Terry A., date unknown. *A day at Cottonwood Spring: Joshua Tree National Monument trail guide.*

Powell, R.E., 1980. *Geology of the crystalline basement complex, eastern Transverse Ranges, Southern California*: Ph.D. dissertation, California Institute of Technology.

Powell, R.E., 1982, Crystalline basement terranes in the southern eastern transverse ranges, California: *in* Cooper, J.D. (ed.) Geologic excursions in the Transverse Ranges: guidebook prepared for the 78th annual meeting of the Cordilleran section of the Geological Society of America, Anaheim, California, April 19-21.

Ruff, Robert W., Mark E. Unruh and Paul A. Bogseth, 1982. *Mineral resources of the eastern Transverse Ranges of Southern California, in* Fife, D.L., and Minch, John A., editors, Geology and mineral resources of the California Transverse Ranges: Santa Ana. South Coast Geological Society, p. 222-249.

Saul, R.B., Gray, C.H., Jr., and Evans, J.R., 1961. *Riverside County mines and minerals*: California Division of Mines and Geology, open-file report (unpublished).

Trent, D.D., 1998. *Mines in Joshua Tree National Park*; California Geology, vol. 51, no. 5, October, p. 17.

Tucker, W.B., and Sampson, R.J., 1945. *Mineral resources of Riverside County*: California Journal of Mines and Geology, vol. 41, no. 3, p. 121-144.

Weathering combined with strong vertical jointing in this granitic body produced these vertical slabs which appear to be sandwiched together. Note the small round spheroids – the result of continued weathering of granitic blocks (photo by John D. Clark. 2004)

HISTORIC MINING EQUIPMENT AND PROCESSES WITHIN JOSHUA TREE NATIONAL PARK

Margaret R. Eggers, PhD, *Eggers Environmental, Inc.,*
Oceanside, CA 92054
D.D. "Dee" Trent, PhD, *Professor Emeritus,*
Citrus College, Glendora, CA 91741

INTRODUCTION

Joshua Tree National Park has a rich diversity of historic mine and mill sites within its boundaries. This article provides a description of the processes and equipment used to process gold ore in the area. Interestingly, the most economically successful, mining-related operation within the Park was not a gold mine, but the ore processing operation which extracted gold for the miners. Bill Keys and his son ran the Wall Street Mill for decades, processing ore for many of the miners within the Park area. The Wall Street Mill can be reached by walking a little over a mile from the small parking area just east of the Barker Dam parking lot.

This legacy of ore mining and processing has left its mark in the form of mine shafts, pits, tailings, and the decaying remains of equipment used to mine and process the gold ore. Most mines within the Park area were rather small operations. Therefore, you won't see remains of large mining and milling operations such as the one at the ghost town of Bodie, north of Mono Lake.

The basic equipment used in ore processing within Joshua Tree is essentially the same as that operated at mine sites in California and throughout the West. The Wall Street Mill, recognized on the National Register of Historic Places, is a well preserved and fairly complete example of a small gold processing mill using late 19th century processes. One can stand outside the Wall Street Mill and imagine what it must have been like to operate this equipment many decades ago. To help you interpret the ore processing equipment you may see during your wanderings in Joshua Tree, we have assembled diagrams from the same era as the equipment you will see and provided explanations on how this equipment was used and how common gold extraction processes worked. The processes described here can be read about in greater detail in historic ore processing texts (Simons, 1924 and Richards et al., 1925). As you wander around the old mine sites both in this park and at any historic site, please remember that mine sites are inherently dangerous due to unstable slopes, open shafts and the decrepit, decaying remains of mining refuse and equipment. PLEASE - remember the remains of the mining operations you are looking at are historic reminders of our rich heritage – take only pictures – leave only foot prints! Please preserve this historic experience for those who will come after you.

Historic Ore Milling Equipment

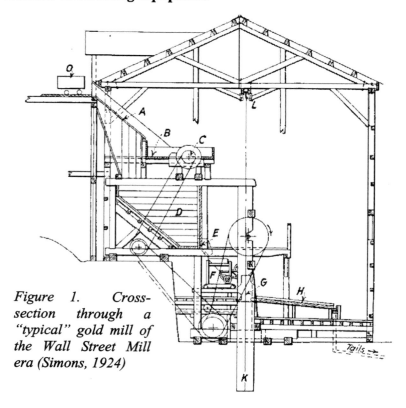

Figure 1. Cross-section through a "typical" gold mill of the Wall Street Mill era (Simons, 1924)

Figure 1 shows a common stamp mill layout which is similar to that of the Wall Street Mill and many other small milling operations common to Southern California. The mill was usually constructed on a hill or incline to allow gravity to feed the ore into the various crushing and refining steps within the mill. Gold ore was delivered into the mill building in small increments in ore carts (**O,** figure 1). The ore was dumped onto a grate set above the ore bin. This grate, called a "grizzly" (**A,** figure 1), was set at an angle above the ore bin. The grizzly (figure 2) separated the finer materials from the large chunks of ore which need to be broken up in a rock crusher or breaker.

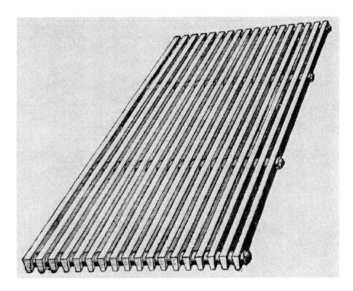

Figure 2. The "Grizzly." This heavy metal screen separated fine from coarse ore fragments. The larger ore chunks slid off the top of the screen and were diverted into a rock crusher or breaker. Grizzlies were usually made of stationary, heavy metal bars at a specific distance apart which controlled the size of the ore which dropped into the ore bin to be fed into the stamp mill (Richards et al., 1925).

Rock Crushers and Breakers

Ore that fell through the grizzly dropped into the ore bin (**D**, Figure 1). The coarser ore that slid off the grizzly collected onto the crusher floor (**B**, Figure 1). This coarse material was then processed through a rock breaker (**C**, Figure 1) which reduced the large ore chunks into finer fragments prior to falling into the ore bin (**D**, Figure 1). The rock breaker, or crusher, (Figure 3) broke up large pieces of ore into smaller fragments by pounding the ore between two metal plates or jaws, one stationary and one set at an angle which is thrust against the stationary plate. The ore to be processed is dropped into the wedge-shaped area between the two plates, referred to as the "mouth," and the crushed ore fell through the "throat" or bottom of the crusher when it reached the desired size. In small milling operations, ore that slid off the grizzly may have been reduced in size manually using a sledge hammer and lots of sweat!

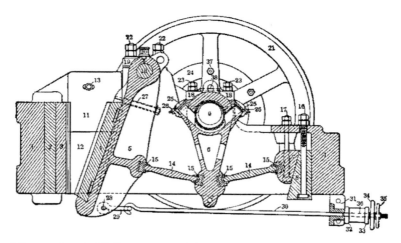

Figure 3. A Blake-type rock breaker. Large ore pieces were crushed between the two metal plates on the left side of the figure. The left plate was stationary while the right plate was set at an angle and crushed ore chunks against the stationary plate (Richards et al., 1925).

Ore that fell through the grizzly or was processed by the breaker collected in the ore bin (**D**, Figure 1). The ore bin usually had enough capacity for the mill to run for 24-hours without additional ore supply in case the rock breaker failed, or ore supply from the mine was interrupted. The ore bin fed by gravity into a feeder (**F**, Figure 1) which was placed at the back of the stamp mortar (**G**, Figure 1). Once the ore reached the inside of the stamp mortar, it was pulverized with water into a fine rock pulp.

Pulverizing the Ore: The Stamp Mill

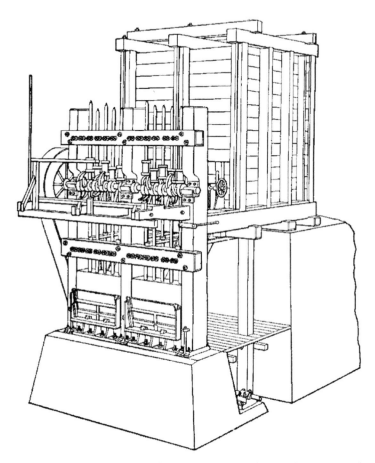

Figure 4. A typical "California Stamp Mill." This particular stamp mill has two sets of five stamps each. The collection of the stamps, mortar, cams and cam shafts and wooden frame together is referred to as a "stamp battery" (Richards et al., 1925).

The stamp mill (**G**, Figure 1; and Figure 4) was the most crucial part of the ore milling process. The type of gravity stamp mill such as those seen within the Park and throughout California, Arizona, Colorado, Montana, Nevada and elsewhere was known as the "California Stamp Mill" (Richards et al., 1925). Examples within the Park include the two-stamp mill at the Wall Street Mill, and the 10-stamp mill at the Lost Horse mine. The Lost Horse mine is located on Lost Horse Mountain and can be reached by a dirt road and hiking trail off of Keys View Road.

Figure 4 shows a schematic of a 10-stamp, California Stamp Mill. One or more sets of two to five heavy metal hammers, or stamps, are lifted and dropped into a metal mortar. The mortar is the large cast-iron box where the stamps pound the ore into rock powder. Each stamp is raised and lowered into the mortar by a cam shaft. Each stamp had its own individual cam which is designed such that each hammer fall was staggered and no two stamps would strike the mortar at the same time. It was thought that if the stamps fell simultaneously, the vibration would be so strong as to gradually shake the stamp mill apart.

Gold Extraction: Amalgamation with Mercury

One common process to release the gold from the ore was based upon the affinity of gold for metallic mercury. When the two metals come into contact with each other, they form an alloy. This gold-mercury alloy is known as a mercury amalgam. As the ore is pulverized with water in the stamp mill, the ore slurry runs out of the mortar onto long, inclined copper plates, called apron plates or amalgamation plates, which were placed in front of the stamp mill battery. As the powdered rock slurry moves over the mercury-coated plates, the gold adheres to the mercury and the rock pulp continues off the amalgamation plate. Figure 5 shows an illustration of a 10-stamp mill with the apron-type amalgamation plates situated in front of the stamp mill.

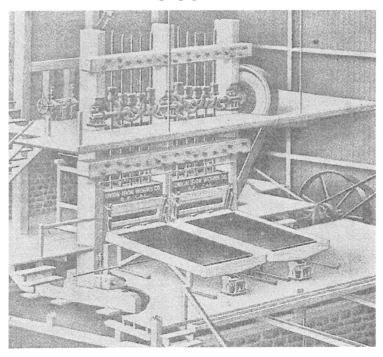

Figure 5. Illustration of the interior of a 10-stamp mill with apron-type amalgamation plates. The slurry of water and pulverized ore produced by the stamps flowed across the mercury-coated apron plates, capturing the gold from the slurry. (Simons, 1924).

The amalgamation plates were typically copper or silver-coated copper. The mercury was applied evenly by sprinkling or brushing the mercury onto the surface of the copper or silver plating until it was distributed evenly. Once enough ore was processed that the amalgam seemed to be reasonably saturated with gold, the mercury amalgam would be removed from the plates and fresh mercury applied (Figure 6). Typically a rubber scraper would be used to remove the gold-saturated amalgam from the apron plates. Unless the ore being processed was very rich in gold, the plates would be scraped and re-coated with mercury once or twice a day. Generally, mercury was also added in small amounts to the stamp mortar to facilitate the gold extraction. It took skill and experience to know just how much mercury to apply to the plates, and how much, and when, to add mercury to the stamp mortar, and how often to remove the mercury amalgam and re-apply fresh mercury to the plates.

Figure 6. 1912 photograph taken in the Yellow Aster mine, Randsburg Mining District, CA. This 100-stamp mill was one of the largest operating in the area. Here workers are in the process of scraping gold-saturated amalgam from apron plates situated in front of each stamp battery (collection of John W. Robinson, Guerin-Place et al., 1998).

When the mercury amalgam was removed from the plates, it was placed in a small furnace or retort (Figure 7) in small batches. The inside of the pot was coated with a paste of clay and graphite mixture to keep the gold residue (sometimes referred to as a "gold sponge") from sticking to the inside of the retort pot. The pot would be filled about two-thirds full and heated over an assayers furnace or blacksmith's forge. As the amalgam is heated, the mercury boils off and condenses in the pipe or coil. The mouth of the coil is placed just at the surface of a pot of water to help condense and collect the mercury. Since the pot is heated to a temperature hot enough to volatilize the mercury, but not the gold or other metals, the gold is left behind in the pot. The condensed mercury could be reused again and again. This crude refined gold product could then be further refined at the mill or be shipped to a smelter or mint for additional refining to separate out metallic impurities from the gold.

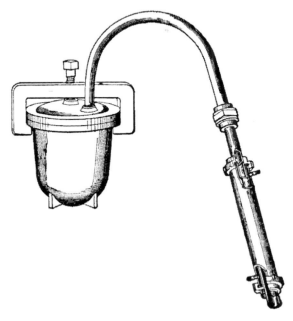

Figure 7. Pot retort. Gold/mercury amalgam is heated to condense off mercury and concentrate gold and other metallic minerals for further refining (Richards et al., 1925).

Ore Concentration Table: Further Separation of Gold From the Ore

While mercury amalgamation was a good way to extract much of the free gold from the ore, it was not even close to being 100% effective in removing all the gold. Some ore particles may have been too coarse to release gold to the mercury, and some other ore minerals such as sulfides could sour the amalgamation effect, hindering the gold from binding with the mercury. For this reason, the pulverized rock slurry coming off of the amalgamation table would be further processed using some type of concentration table.

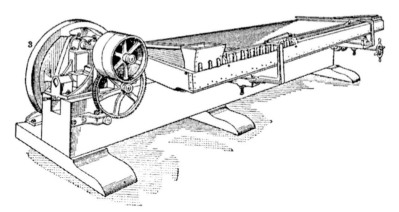

Figure 8. Diagram of a Wilfley-type concentration table (Richards et al., 1909).

Although there were many kinds and variations of concentration tables, one of the most common was known as the Wilfley table, named after its inventor, Arthur R. Wilfley. Figure 8 is an illustration of one type of Wilfley concentration table. Whatever the variety, all concentration tables operated on the principle that materials will sort themselves according to their size and or density if kept mobile in water and agitated slightly. As the rock slurry leaves the amalgamation table, it is directed to the concentration table. The concentration table is constructed with an impervious deck, usually a wood base covered by linoleum. Small strips of wood called "riffles" are tacked onto the table deck in varying arrangements but typically with shorter strips at the side where the slurry feeds onto the table, and longer strips at the bottom of the table. The riffles help concentrate the denser grains of the ore which still have a metallic content. The agitation gradually moves the sandy, quartz-rich fraction toward the other end of the table which is taken off to the tailings pile (Figure 9).

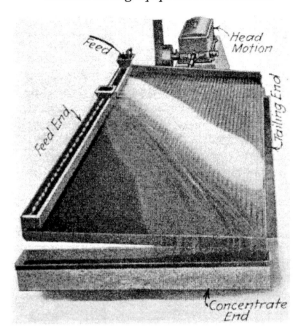

Figure 9. Illustration of a concentration table. The slurry coming from the amalgamation table feeds onto the table on the left side and the table is agitated by a motor on the far end of the table. The fraction of the material which still contains the denser, metallic minerals (the concentrate) is separated by gravity to the near end of the table, and the tailings which should be essentially free of any economical gold content, is moved to the right side of the table to be disposed of in the tailings pile (Simons, 1924).

The Wall Street Mill

The Wall Street Mill was installed in 1932 or 1933 (Hickman, 1977) and still exists today much as it was when Willis Keys stopped processing ore in the mid-1960s (Figure 10). The excellent state of preservation of the mill and its small size make it a good place to develop am understanding of the mill layout and ore processes described in this article. The photograph shows the mill's condition in the summer of 2004.

Figure 10 (facing page). Interior of the Wall Street Mill as it appears today. The two-stamp mill is seen at the rear of the photograph in the upper right. One stamp is visible in the opening of the stamp battery itself. The inclined table in front of the stamp mill in the center-right of the photo is the amalgamation table that was once covered with copper sheeting which would have been coated with mercury. The slurry of water and pulverized rock flowed across the amalgamation plate. As the rock slurry reached the end of the amalgamation plate, the slurry was directed to a Wilfley-type concentration table (foreground, left) to further remove gold-rich particles from the pulverized rock. Some of the wooden "riffles" are still attached to the concentration table. At the lower end of the concentration table, tailings were diverted to one side, and ore which still contained gold would be separated out for further processing (photo by M.R. Eggers, 2004).

Figure 11 and Table 1 illustrate the locations of mines in the Joshua Tree area between 1930 and 1943. As noted on the map, some of these mines are known to have brought their ore to Bill Keys for processing at the Wall Street Mill. Most mining activity during this time period was focused on re-working existing claims rather than establishing new mines. The advent of the use of cyanide extraction procedures also made it economical to process mine tailings of earlier operations (Hickman, 1977).

Table 1. Mines operating between 1930 and 1943 (Figure 11)

Map No.	Mine Name	Map No.	Mine name	Map No.	Mine name	Map No.	Mine Name
1	Hidden Gold	11	Gold Park\ Morton's Mill	21	Lureman	31	Yellow Jacket/ Desert Star/ Iron Chief/ Black Eagle
2	Black Warrior	12	Brooklyn	22	McKeith		
3	Gold Point	13	Supply	23	Star	32	Snow Cloud
4	Gold Fields of America	14	Desert King	24	Virginia Dale	33	Sunrise/ Grassy Hills/ Rainbow Hole
5	Paymaster	15	Lost Horse/ D.C.	25	Golden Rod Group	34	Coyote Lode
6	Gold Crown	16	Desert Queen	26	Heeley & Cross	35	Mastodon
7	O.K./ Top Nest	17	Jupiter	27	Meek Group	36	Southern Cross Lode
8	Ivanhoe	18	Big Boze	28	Gold Coin	37	Yucca Butte/ Jim Reed/
9	Golden Bee/ Dicky Boy/Mabel	19	Frank Hill	29	Blue Bell		Diane/ Mystery/ Mule Shoe
10	Carlyle	20	Lena	30	El Dorado	38	Gypsy

Figure 11. Mines operating between 1930 and 1943, The circled mines are those that Bill Keys milled for during that time period. See Table 1 for mine identification (Hickman, 1977)

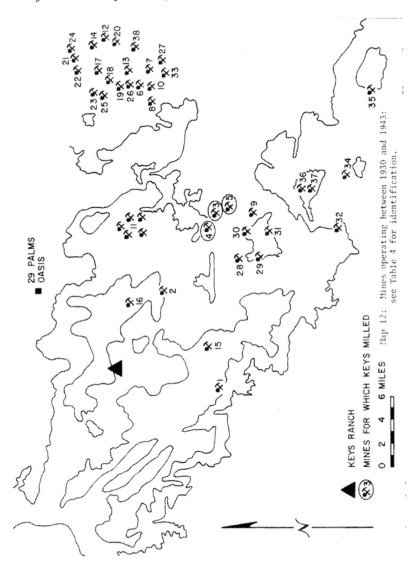

REFERENCES AND ADDITIONAL READING

Guerin-Place, Rosemary, Roefs, Lyn, and Twomey, Kathleen, 1988. *California Gold Mines: A Sesquicentennial Photograph Collection.* California Geological Survey Publication, CD-98-001.

Hickman, Patricia Parker, 1977. *Country Nodes: An Anthropological Evaluation of William Keys' Desert Queen Ranch, Joshua Tree National Monument, California,* Western Archeological Center, Publications in Anthropology No. 7, Tucson AZ, National Park Service, U.S. Department of the Interior.

Richards, Robert H., Bardwell, Earl S., and Goodwin, Edwin G., 1909. *A Text Book of Ore Dressing, 1st Edition.* McGraw-Hill Book Company, New York.

Richards, Robert H., Locke, Charles E., and Bray, John L., 1925. *A Text Book of Ore Dressing, 2nd Edition.* McGraw-Hill Book Company, New York.

Sagstetter, Beth, and Sagstetter, Bill, 1998. *The Mining Camps Speak: A New Way to Explore the Ghost Towns of the American West.* BenchMark Publishing of Colorado.

Simons, Theodore, 1924. *Ore Dressing Principles and Practice, 1st Edition.* McGraw-Hill Book Company, New York.

CRUST FORMATION AND EVOLUTION IN SOUTHERN CALIFORNIA: FIELD AND GEOCHRONOLOGIC PERSPECTIVES FROM JOSHUA TREE NATIONAL PARK

Andrew P. Barth, PhD, *Department of Geology, Indiana University ~Purdue University, Indianapolis*
Joseph L. Wooden, PhD, *U.S. Geological Survey, Menlo Park*
Drew S. Coleman, PhD, *Department of Geological Sciences, University of North Carolina*
Janet L. Jarvis, *Department of Geology, Indiana University ~Purdue University, Indianapolis*

INTRODUCTION

Joshua Tree National Park is an excellent natural laboratory for examining the timing and tectonic processes associated with continental crust formation in the Laurentian craton and crustal evolution at the North American plate margin. Metamorphic rocks in the Park are probably the largest semi-contiguous exposure of Precambrian basement in the far western United States, and originally lay at the southwestern edge of Laurentia at the time of rifting and formation of the Cordilleran miogeocline and the Pacific basin. Joshua Tree includes the majority of the eastern Transverse Ranges of California, and as such includes rocks extending more than 100 kilometers across the strike of the Mesozoic Cordilleran orogen. In addition, mountain ranges and basins of the Park were shaped in the Cenozoic at the intersection of the eastern California shear zone and the southern San Andreas fault system.

In this short contribution we summarize ongoing field and geochronologic studies aimed at constructing a chronologic framework for examining tectonic processes of crust formation and crustal evolution in this key region of southern California. Our work incorporates and builds upon previous studies by Miller (1946), Rogers (1954), Dibblee (1967, 1968), Hope (1969), Powell (1981), Calzia (1982), Stone and Pelka (1989), and Howard (2002). These earlier studies and our new mapping and geochronologic database are summarized in the simplified bedrock map of the Park shown in Figure 1. Below we summarize the characteristics of the major rock units of the park, existing understanding of their relationships, and preliminary U-Pb zircon and monazite estimates of their ages.

PALEOPROTEROZOIC BASEMENT ROCKS

Continental crustal rocks in southwestern North America formed in Precambrian time. In Arizona, Colorado and New Mexico, this crust formed about 1760-1615 Ma directly from the underlying mantle (Condie, 1982; Karlstrom and Bowring, 1988; Anderson, 1989; DeWitt, 1989; Bowring and Karlstrom, 1990; Hawkins et al., 1996; Duebendorfer et al., 2001). In Southern California substantial continental growth also occurred at this time, but geochemical and geochronologic studies suggest the presence of a cryptic older crustal component, perhaps as old as 2500 Ma (DeWitt et al., 1984; Bennett and DePaolo, 1987; Wooden et al., 1988; Wooden and Miller, 1990; Barth and others, 2000, 2001). Therefore, initial continental crust formation in Southern California may have occurred earlier than areas to the east and north, or may have involved recycling of some older continental crust. Petrogenetic studies of Paleoproterozoic rocks in Jpshua Tree are aimed at understanding the timing and tectonic processes of crust formation and the relative role of recycling in crustal growth.

Paleoproterozoic rocks in the central and western Pinto Mountains, Lost Horse and western Hexie Mountains region are composed of at least six different rock types. The most widespread map unit, biotite gneiss, is a composite unit composed of biotite and/or muscovite, sillimanite, garnet quartzofeldspathic gneiss, biotite and/or muscovite schist, amphibolite, metagabbro and quartzite (Figure 1). Gross compositional layering is defined by interlayering of this biotite gneiss with light tan granitic gneiss, which locally defines tight to isoclinal map-scale folds. Textural variations in the granitic gneiss suggest that it was originally plutonic rocks, but the relationship(s) of the protoliths to the more abundant biotite gneiss unit is uncertain. All rock types in these ranges have undergone moderately high temperature metamorphism, which formed the observed mineral assemblages, approximately synchronous with folding and pervasive ductile deformation. Centimeter to meter scale folding of the planar fabric is common, and these folds typically have hinge lines subparallel to a mineral streak lineation. In the Hexie Mountains and eastward into the Eagle Mountains, similar mica-rich gneisses are interleaved with orthogneisses, which include equigranular to porphyritic granitic rocks in the northeast and generally more strongly deformed augen gneisses to the south and west.

U-Pb zircon ages for orthogneisses in Southern California suggest that discrete magmatic episodes occurred at approximately 1780, 1750, and 1690-1675 Ma. Orthogneisses in Joshua Tree include representatives of the two younger of these three regional magmatic episodes. With few exceptions, these ages are indistinguishable from published ages for the Yavapai and Mazatzal belts to the east (Karlstrom and Bowring, 1993). Paleoproterozoic rocks in Joshua Tree also preserve evidence for thermo-tectonic events as high U/Th rims on detrital or primary igneous zircons and episodic growth of monazite (Coleman et al., 2002 and in prep.). Ages for these minerals demonstrate that metamorphism occurred at approximately 1740 Ma (correlative to an unnamed deformational event in the Yavapai province), 1700 Ma (correlative to the Yavapai orogeny in Arizona and the Ivanpah orogeny in the eastern Mojave province) and 1650 Ma (correlative to the Mazatzal orogeny, now recognized as far north as northern Arizona [Duebendorfer *et al.*, 2001]).

MESOPROTEROZOIC SEDIMENTARY ROCKS

A sequence of weakly deformed and recrystallized sedimentary rocks overlies Paleoproterozoic gneisses and 1693 Ma granite porphyry in the Pinto and Eagle Mountains. This sedimentary cover sequence is >1 km thick and includes basal conglomerate with interbedded quartzite, cross-bedded quartzite, pelitic rocks, laminated Fe-rich rocks, and dolomitic carbonate rocks. Although all exposures examined thus far have been metamorphosed to upper greenschist/lower amphibolite facies and mildly deformed, sedimentary structures are locally well preserved. Planar tabular and planar wedge cross-bedding suggest that sediment sources lay to the north and west of the Pinto Mountains in present coordinates. Paleomagnetic data (Carter et al., 1987) suggests only minor post-depositional clockwise rotation of this region.

We informally call these rocks the Pinto Mountain Group, after the part of the outcrop belt which is best exposed and least deformed, and to avoid implied correlations attached to earlier nomenclature. Detrital zircons in four quartzite samples are predominantly 1650 – 1800 Ma, but populations as old as 3400 Ma are also present. This age distribution suggests that the maximum depositional age of these rocks is about 1630 Ma, but the lack of any younger Mesoproterozoic grains suggests that the rocks are no younger than about 1450 Ma. Our preliminary mapping and detrital zircon geochronology therefore indicate that the Pinto Mountain Group is younger than originally suggested by Powell (1981), yet probably too old to be correlated with the Big Bear Group as suggested by Cameron (1981). As an alternative, we propose that these rocks represent a pre-1.4 Ga part of Succession A, broadly equivalent to the Muskwa Group of northern British Columbia and the Wernecke and Belt supergroups of the southern Canadian and northern U.S. Cordillera (Ross and others, 2001), and/or the upper part of the Tonto Basin supergroup in Arizona (Cox and others, 2002).

MESOPROTEROZOIC INTRUSIVE ROCKS

Mesoperthitic monzonite, granodiorite and granite, locally characterized by clots of biotite and hornblende, intrude Paleoproterozoic gneisses in the eastern Eagle Mountains (Powell, 1981). The mafic mineral clots are interpreted to represent alteration of pyroxene, in which case the monzonitic rocks are equivalent to mangerite. Powell (1981, 1993) correlated these rocks to the San Gabriel syenite complex, dated at 1191 ± 4 Ma (Barth and others, 1995, 2001). This correlation is consistent with a 1198 ± 8 Ma age we recently measured on zircons from a quartz monzonite from this unit.

MESOZOIC SEDIMENTARY ROCKS

Plutonic rocks in the Coxcomb Mountains intrude sedimentary rocks correlated with the McCoy Mountains Formation (Calzia, 1982; Stone and Pelka, 1989). The McCoy Mountains Formation is a dominantly fluvial sequence more than 7 km thick resting on volcanic and volcaniclastic rocks of the Middle to Late Jurassic Dome Rock Sequence, which in turn lie in depositional contact above Paleozoic and early Mesozoic clastic and carbonate rocks. The top of the McCoy Mountains Formation is nowhere exposed.

The McCoy Mountains Formation is a siliciclastic sequence recording the transition from a cratonal depositional setting to an arc unroofing sequence. The timing of deposition of this sequence has been controversial, and hence the tectonic implications of the formation and infilling of such a deep continental basin has been uncertain. Recent work on populations of detrital zircons in the type area of the formation, east of Joshua Tree, yielded maximum depositional ages that decrease from 116 Ma near the base to 84 Ma near the top of the section (Barth and others, 2004). The detrital zircon data suggest that most of the formation was deposited between mid-Early and mid-Late Cretaceous time, and represents a retro-arc foreland basin, deposited behind the active, evolving Cretaceous Cordilleran continental margin magmatic arc that lay to the west, and in the foreland of the actively deforming Cretaceous Sevier and Maria fold and thrust belts.

MESOZOIC PLUTONIC ROCKS

The elongate east-west form of Joshua Tree, and the eastern Transverse Ranges encompassed within it, places the Park astride the strike of Mesozoic magmatic arcs of the southern Cordillera. Our work on the variety of arc plutonic rocks in Joshua Tree is an effort to build upon and join together studies of individual igneous bodies by Brand (1985), James (1989), Hopson (1996) and Mayo and others (1998), and proposed regional correlations developed by Powell (1981).

Geochemical and U-Pb geochronologic data suggest that plutonic rocks in Joshua Tree can be grouped into suites comprising segments of four temporally discrete Mesozoic magmatic arcs. Quartz monzonitic to granitic rocks, locally porphyritic, are of Early Triassic age and are exposed through the central part of the park, including porphyritic plutonic rocks exposed at Indian Cove and in central Music Valley. These plutons are part of a Permo-Triassic arc which marks the initiation of convergent margin magmatism in the southern U.S. Cordillera (Barth and others, 1997). Monzodioritic to granitic rocks of Middle Jurassic age are extensively exposed in a northwest-trending belt through the central and eastern Pinto Mountains and the northern Eagle Mountains, and are well exposed in the Gold Park and upper Music Valley areas. These plutonic rocks are part of the Kitt Peak – Trigo Peaks superunit, which extends southeastward into Arizona and northern Sonora (Tosdal and others, 1989). Granites of Late Jurassic age are exposed in a northwest-trending belt west of, and subparallel to this belt of Middle Jurassic rocks; these younger arc plutonic rocks include the White Tank granite and granite exposed in the Cottonwood area. Granodioritic to granitic rocks of Late Cretaceous age are exposed in the western part of the park, and include the granodiorite of Porcupine Wash in the southern Pinto Basin, the granite of Smoke Tree Wash, and the Palms granite in Hidden Valley – Lost Horse Valley.

Dikes and dike swarms with a wide range of compositions and textures cut plutons throughout the park, but are particularly abundant in the Eagle and eastern Pinto Mountains. These dikes are at least in part correlative with the older part of the Independence swarm described to the north in the Sierra Nevada and Mojave Desert (Ron and Nur, 1996; Coleman and others, 2000; Carl and Glazner, 2002); James (1989) presented a strongly discordant 147 Ma U-Pb zircon age for a porphyritic rhyolite dike in Big Wash in the Eagle Mountains.

MESOZOIC SHEETED COMPLEX

Mesozoic intrusive rocks display a significantly different style of emplacement in the Little San Bernardino Mountains, in the westernmost part of Joshua Tree. Plutonic rocks, contemporaneous and broadly compositionally similar to Jurassic and Cretaceous plutons described above, intruded Proterozoic gneisses as concordant to slightly discordant meter to decimeter thick sheets. Where these rocks have been mapped, the sheets typically dip gently to moderately north and east. Preliminary thermobarometric results suggest that some plutonic components of the complex crystallized at higher pressures than are typical of plutons exposed to the east in the park, suggesting that the Mesozoic sheeted complex represents exposure of the mid-crustal roots of the discordant upper crustal plutons (Wooden and others, 1994, 2001).

LARAMIDE EXHUMATION

Powell (1981) developed a comprehensive working hypothesis for the structural evolution of the eastern Transverse

Ranges, based on the concept of suspect (or exotic) tectonostratigraphic terranes. Since that time, several studies have identified multiple geochronologic and petrologic ties between inferred suspect terranes, which imply that these terranes need not be viewed as exotic (Bennett and DePaolo, 1987; Anderson and others, 1992; Bender and others, 1993; Barth and others, 2001). We hypothesize that the large-scale distribution of rock units across Joshua Tree results principally from differential Laramide (and perhaps post-Laramide) exhumation.

Proterozoic basement rocks in the Park can be divided into two provinces, western amphibolite facies paragneiss and orthogneiss, and eastern granite porphyry and orthogneiss overlain by typically weakly deformed greenschist to amphibolite facies metasedimentary rocks of the Pinto Mountain Group. These divisions correspond in some senses to the "San Gabriel terrane" and "Joshua Tree terrane" of Powell (1981) and correspond closely to the "Hexie Mountains assemblage" and "Eagle Mountains assemblage" of Powell (1993). However, Bender and others (1993) showed that Paleoproterozoic orthogneisses of both provinces are compositionally similar, and Powell (1993) suggested that rocks characteristic of both were intruded by Mesoproterozoic mangerite. Furthermore, geochronologic data indicate that deposition of the Pinto Mountain Group was postorogenic with respect to plutonism, regional metamorphism and deformation of the Paleoproterozoic gneisses. The boundary between these two provinces is locally intruded by Mesozoic plutonic rocks, but in at least two localities is inferred to be a brittle east-dipping normal fault of Mesozoic or Cenozoic age (Postlethwaite, 1988; Howard, 2002). We agree with Powell (1993) that the boundary was originally marked by one or more faults; the many shared characteristics between these provinces suggest that they are not exotic terrane but represent distinct structural levels within a single coherent Proterozoic crustal section, a part of the greater Mojave crustal province of Wooden and others (1988).

The distribution of Mesozoic rocks across Joshua Tree is consistent with this hypothesis. The central and eastern parts of the Park are characterized by Mesozoic plutons with generally sharp intrusive contacts and low-grade recrystallization of wallrocks. Crystallization pressures of these plutons suggest emplacement at depths of 12 km or less (Anderson and others, 1992). Preservation of Cretaceous sedimentary rocks in the eastern part of the Park also suggests relatively limited Laramide and younger exhumation there. In contrast, the western part of the Park is a transition to the Mesozoic sheeted complex, characterized by intimate interlayering and generally more concordant contacts between intrusive rocks and Paleoproterozoic gneisses. Jurassic and Cretaceous intrusive rocks in the sheeted complex crystallized at pressures of 4.5 to 6 kb, consistent with exhumation of the western part of the Park from depths of 16 to 22 km (Wooden and others, 1994, 2001). Exposures of Orocopia metagraywacke immediately south of the Park's western margin, and the likely occurrence of correlative rocks to the west prior to initiation of dextral slip on the southern San Andreas fault system, implicates differential Laramide exhumation as the cause of the observed east to west increase in exhumation.

CENOZOIC FAULTING AND BASALT VOLCANISM

Previous work (Stull and McMillan, 1973; Carter et al., 1987) and our own preliminary petrologic and geochronologic data suggest that basaltic volcanism occurred in proximity to sinistral faults that bound or cut across Joshua Tree, and that volcanism resulted from partial melting in the asthenosphere synchronous with transrotational faulting in the lithosphere in the eastern Transverse Ranges. At least ten discrete basalt centers are present on or near to fault blocks inferred to have undergone clockwise rotation associated with sinistral faulting in the eastern Transverse Ranges, and many centers occur in close proximity to block boundaries (Powell, 1993; Figure 1). Western centers include dikes and columnar-jointed stocks, locally bearing ultramafic and feldspathic xenoliths and xenocrysts. Other centers typically include one to eight flows with dense cores and vesicular flow tops. A better understanding of the petrogenesis of these basaltic rocks could provide insight into the composition and thermal state of the asthenosphere during initiation of the southern San Andreas fault system, as well as determining the timing of block rotation inferred from declination anomalies. The basalts are described in more detail by Probst et al. (2004, this volume).

ACKNOWLEDGEMENTS

Research in Joshua Tree National Park was supported by National Science Foundation grants EAR-9614499, 9614511, and 0106881, National Geographic Society grant 7214-02, and by continuing generous support from the National Park Service and the Joshua Tree National Park Association. We appreciate enthusiastic field assistance from Kenneth Brown, Kristin Hughes, Leda Jackson, Nick Kaufman, Sarah Needy, Derrick Newkirk, Emerson Palmer, Kelly Probst, and Joshua Richards.

Editor's note: Dr. Rick Hazlett, Associate Professor in the Geology Department for Pomona College provided technical review for this paper. Rick notes that Dr. Barth's ongoing work is providing us with not only better defined geologic mapping in the western portions of the park, but also a fresh understanding of the role and timing of faulting in the area. This is a great example of how our work as geologists builds on earlier interpretations, constantly giving us a fresh perspective and bringing us ever closer to a more complete understanding of the landscape around us.

REFERENCES

Anderson, J.L., Barth, A.P., Young, E.D., Bender, E.E., Davis, M.J., Farber, D.L., Hayes, E.M., and Johnson, K.A., 1992. *Plutonism Across the Tujunga-North American Terrane Boundary: A Middle To Upper Crustal View of Two Juxtaposed Magmatic Arcs, in* Bartholomew, M.J., Hyndman, D.W., Mogk, D.W., and Mason, R. (eds.), Basement Tectonics 8: Characterization and Comparison of Ancient and Mesozoic Continental Margins - Proceedings of the Eighth International Conference on Basement Tectonics, Kluwer, Dordrecht, 205-230.

Anderson, P., 1989. *Proterozoic Plate Tectonic Evolution of Arizona*, Arizona Geological Society Digest 17, 17-55.

Barth, A.P., Wooden, J.L., Tosdal, R.M., Morrison, J., Dawson, D.L., and Hernly, B.M., 1995. *Origin of Gneisses in the Aureole of the San Gabriel Anorthosite Complex, And Implications For the Proterozoic Crustal Evolution of Southern California*, Tectonics 14, 736-752.

Barth, A.P., Tosdal, R.M., Wooden, J.L., Howard, K.A., 1997. *Triassic Plutonism in Southern California: Southward Younging of Arc Initiation Along a Truncated Continental Margin.* Tectonics 16, 290-304.

Barth, A.P., Wooden, J.L., Coleman, D.S., and Fanning, C.M., 2000. *Geochronology of the Proterozoic Basement of Southwesternmost North America, And the Origin And Evolution of the Mojave Crustal Province,* Tectonics 19, 616-629.

Barth, A.P., Wooden, J.L., and Coleman, D.S., 2001. *SHRIMP-RG U-Pb Zircon Geochronology of Mesoproterozoic Metamorphism And Plutonism in the Southwesternmost United States,* Journal of Geology 109, 319-327.

Barth, A.P., Wooden, J.L., Jacobson, C.E., and Probst, K., 2004. *U-Pb Geochronology And Geochemistry of the McCoy Mountains Formation, Southeastern California: A Cretaceous Retroarc Foreland Basin,* Geological Society of America Bulletin 116, 142-153.

Bender, E.E., Morrison, J., Anderson, J.L., and Wooden, J.L., 1993. *Early Proterozoic Ties Between Two Suspect Terranes And the Mojave Crustal Block of the Southwestern United States,* Journal of Geology 101, 715-728.

Bennett, V.C., and DePaolo, D.J., 1987. *Proterozoic Crustal History of the Western United States As Determined By Neodymium Isotopic Mapping,* Geological Society of America Bulletin 99, 674-685.

Bowring, S.A., and Karlstrom, K.E., 1990. *Growth, Stabilization, And Reactivation of Proterozoic Lithosphere in the Southwestern United States,* Geology 18, 1203-1206.

Brand, J.H., 1985. *Mesozoic Alkalic Quartz Monzonite And Peraluminous Monzogranites of the Northern Portion of Joshua Tree National Monument, Southern California,* M.S. thesis, University of Southern California, Los Angeles, 187 pp.

Calzia, J.P., 1982. *Geology of Granodiorite in the Coxcomb Mountains, Southeastern California, in* Frost, E.G., and Martin, D.L. (eds.), Mesozoic – Cenozoic Tectonic Evolution of the Colorado River Region, California, Arizona, and Nevada: San Diego, California, Cordilleran Publishers, 173-180.

Cameron, C.S., 1981. *Geology of the Sugarloaf And Delamar Mountain Areas, San Bernardino Mountains, California*, Ph.D. dissertation, Cambridge, Massachusetts Institute of Technology, 399 pp.

Carl, B.S., and Glazner, A.F., 2002. *Present Extent of the Independence Dike Swarm*, Geological Society of America Memoir 195, 117-130.

Carter, J.N., Luyendyk, B.P., and Terres, R.R., 1987. *Neogene Clockwise Tectonic Rotation of the Eastern Transverse Ranges, California, Suggested By Paleomagnetic Vectors*, Geological Society of America Bulletin 98, 199-206.

Coleman, D.S., Carl, B.S., Glazner, A.F., and Bartley, J.M., 2000. *Cretaceous Dikes Within the Jurassic Independence Dike Swarm in Eastern California*, Geological Society of America Bulletin 112, 504-511.

Coleman, D.S., Barth, A.P., and Wooden, J.L., 2002. *Early To Middle Proterozoic Construction of the Mojave Province, Southwestern United States*, Gondwana Research 5, 75-78.

Condie, K.C., 1982. *Plate-Tectonics Model For Proterozoic Continental Accretion in the Southwestern United States*, Geology 10, 37-42.

Cox, R., Martin, M.W., Comstock, J.C., Dickerson, L.S., Ekstrom, I.L., and Sammons, J.H., 2002. *Sedimentology, Stratigraphy, And Geochronology of the Proterozoic Mazatzal Group, Central Arizona*, Geological Society of America Bulletin 114, 1535-1549.

DeWitt, E., 1989. *Geochemistry And Tectonic Polarity of Early Proterozoic (1700-1750 Ma) Plutonic Rocks, North-Central Arizona*, Arizona Geological Society Digest 17, 149-167.

DeWitt, E., Armstrong, R.L., Sutter, J.F., and Zartman, R.E., 1984. *U-Th-Pb, Rb-Sr, and Ar-Ar Mineral And Whole Rock Isotopic Systematics in a Metamorphosed Granitic Terrane, Southeastern California,* Geological Society of America Bulletin 95, 723-739.

Dibblee, T.W., 1967. *Geologic Map of the Joshua Tree Quadrangle, San Bernardino And Riverside Counties, California,* U.S. Geological Survey Map I-516.

Dibblee, T.W., 1968. *Geologic Map of the Twentynine Palms Quadrangle, San Bernardino And Riverside Counties, California,* U.S. Geological Survey Map I-561.

Duebendorfer, E.M., Chamberlain, K.R., and Jones, C.S., 2001. *Paleoproterozoic Tectonic History of the Cerbat Mountains, Northwestern Arizona: Implication For Crustal Assembly in the Southwestern United States,* Geological Society of America Bulletin 113, 575-590.

Hawkins, D.P., Bowring, S.A., Ilg, B.R., Karlstrom, K.E., and Williams, M.L., 1996. *U-Pb Geochronologic Constraints On the Paleoproterozoic Crustal Evolution of the Upper Granite Gorge, Grand Canyon, Arizona,* Geological Society of America Bulletin 108, 1167-1181.

Hope, R.A., 1969. *the Blue Cut Fault, Southeastern California,* US Geological Survey Professional Paper 650-D, D116-D121.

Hopson, R.F., 1996. *Basement Rock Geology And Tectonics of the Pinto Mountain Fault, San Bernardino County, Southern California,* M.S. thesis, California State University, Los Angeles, 134 pp.

Howard, K.A., 2002. *Geologic Map of the Sheep Hole Mountains 30' X 60' Quadrangle, San Bernardino And Riverside Counties, California,* U.S. Geological Survey Map MF-2344.

James, E.W., 1989. *Southern Extension of the Independence Dike Swarm of Eastern California,* Geology 17, 587-590.

Karlstrom, K.E., Bowring, S.A., 1988. *Early Proterozoic Assembly of Tectonostratigraphic Terranes in Southwestern North America,* Journal of Geology 96, 561-576.

Mayo, D.P., Anderson, J.L., and Wooden, J.L., 1998. *Isotopic Constraints On the Petrogenesis of Jurassic Plutons, Southeastern California*, International Geology Review 40, 421-442.

Miller, W.J., 1946. *Crystalline Rocks of Southern California*, Geological Society of America Bulletin 57, 457-542.

Postlethwaite, C.E., 1988. the *Structural Geology of the Red Cloud Thrust System, Southern Eastern Transverse Ranges, California*, Ph.D. dissertation, Iowa State University, Ames, 135 p.

Powell, R.E., 1981. *Geology of the Crystalline Basement Complex, Eastern Transverse Ranges, Southern California*, Ph.D. dissertation, California Institute of Technology, 441 pp.

Powell, R.E., 1993. *Balanced Palinspastic Reconstruction of Pre-Late Cenozoic Paleogeology, Southern California: Geologic And Kinematic Constraints On Evolution of the San Andreas Fault System, in* Powell, R.E., Weldon, R.J., and Matti, J.C., eds., the San Andreas Fault System: Geological Society of America Memoir 178, 1-106.

Probst, Kelly P., Barth, Andrew P., and Yi, Keewook, 2004. *Neogene Volcanism in Joshua Tree National Park, in*: Mining History and Geology of Joshua Tree National Park, San Diego Association of Geologists Annual Field Guide, 2004., Margaret R. Eggers, editor. Sunbelt Publishing, San Diego California.

Rogers, J.J.W., 1954. *Geology of a Portion of Joshua Tree National Monument, Riverside County*, California Division of Mines and Geology Bulletin 170, Map Sheet 24.

Ron, H., and Nur, A., 1996. *Vertical Axis Rotations in the Mojave: Evidence From the Independence Dike Swarm*, Geology 24, 973-976.

Ross, G.M., Villeneuve, M.E., and Theriault, R.J., 2001. *Isotopic Provenance of the Lower Muskwa Assemblage (Mesoproterozoic, Rocky Mountains, British Columbia), New Clues To Correlation And Source Areas*, Precambrian Research 111, 57-77.

Stone, P., and Pelka, G.J., 1989. *Geologic Map of the Palen-McCoy Wilderness Study Area, And Vicinity, Riverside County, California,* United States Geological Survey Map MF2092.

Stull, R.J., and McMillan, K., 1973. *Origin of Lherzolite Inclusions in the Malapai Hill Basalt, Joshua Tree National Monument, California,* Geological Society of America Bulletin 84, 2343-2350.

Tosdal, R.M., Haxel, G.B., and Wright, J.E., 1989. *Jurassic Geology of the Sonoran Desert Region, Southern Arizona, Southeast California, And Northernmost Sonora: Construction of a Continental Margin Magmatic Arc,* Arizona Geological Society Digest 17, 397-434.

Wooden, J.L., and Miller, D.M., 1990. *Chronologic And Isotopic Framework For Early Proterozoic Crustal Evolution in the Eastern Mojave Desert Region, SE California,* Journal of Geophysical Research 95, 20,133-20,146.

Wooden, J.L., Stacey, J.S., Howard, K.A., Doe, B.R., and Miller, D.M., 1988. *Pb Isotopic Evidence For the Formation of Proterozoic Crust in the Southwestern United States, in:* Ernst, W.G. (ed.), Metamorphism and Crustal Evolution of the Western United States: Englewood Cliffs, Prentice-Hall, 68-86.

Wooden, J.L., Tosdal, R.M., Howard, K. A., Powell, R.E., Matti, J.C., and Barth, A.P., 1994. *Mesozoic Intrusive History of Parts of the Eastern Transverse Ranges, California: Preliminary U-Pb Zircon Results,* GSA Abstracts with Programs 26, no. 2, p. 104.

Wooden, J.L., Fleck, R.J., Matti, J.C., Powell, R.E., and Barth, A.P., 2001. *Late Cretaceous Intrusive Migmatites of the Little San Bernardino Mountains, California,* GSA Abstracts with Programs 33, no. 3, p. A65.

Figure 1. (color foldout, back of guide)

Simplified bedrock geologic map of Joshua Tree National Park, based on Landsat thematic mapper image analysis and field mapping, with additional data from Rogers (1954), Dibblee (1967, 1968), Hope (1969), Powell (1981), and Howard (2002).

NEOGENE VOLCANISM IN JOSHUA TREE NATIONAL PARK

Kelly R. Probst, *Department of Geological Sciences, University of Florida, Gainesville, FL*
Andrew P. Barth, PhD, *Department of Geology, Indiana University ~Purdue University, Indianapolis, IN*
Keewook Yi, *Oral Health Research Institute, Indiana University ~Purdue University, Indianapolis, IN*

INTRODUCTION

The form and variety of the landscape of Joshua Tree National Park is a key to the park's visual identity. The overall form of the landscape, including the elevation difference between the Mojave Desert and Sonoran Desert biomes, was strongly influenced by Neogene strike slip faulting, which is associated in time and space with mafic volcanism (Hope, 1969; Powell, 1981). Stull and McMillan (1973) showed that one such well known locality of mafic volcanic rock in the park, Malapai Hill, is composed of silica-undersaturated basalt which contains ultramafic xenoliths. In this paper, we examine the variety of Neogene volcanic rocks within Joshua Tree. The objective is to understand the interplay between faulting and volcanism within the evolving system of strike slip faults in Southern California.

Geologic Setting

Numerous strike slip faults extend through or are adjacent to Joshua Tree. The most famous of these is the right lateral San Andreas fault, which borders the western edge of the park. In addition there are a series of east-west trending, left lateral faults, including the Pinto Mountain fault along the northern edge of the park, the Blue Cut fault in the central part of the park, and the Substation, Victory Pass, Porcupine Wash and Smoke Tree Wash faults in the southern portion of the Park (Figure 1).

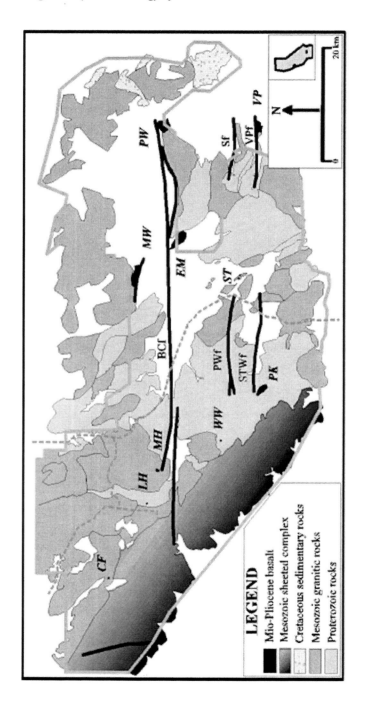

Figure 1. (facing page) Simplified geologic map of Joshua Tree National Park, showing the location of major mapped faults and Neogene volcanic centers. The Blue Cut fault (BCf) extends through the central portion of the Park and has several basalt centers near it, including Lost Horse (LH), Malapai Hill (MH), Mission Well (MW), Eagle Mountain (EM), and Pinto Well (PW). The Victory Pass fault (VPf) and the Porcupine Wash fault (PWf) are associated with the Victory Pass (VP) and Smoke Tree (ST) basalt centers. Other Neogene volcanic centers where studies are still in progress include Covington Flats (CF) in the northwestern part of the park, Washington Wash (WW), and Pinkham Canyon (PK).

The Blue Cut and Victory Pass faults have basaltic volcanic centers associated with them, and are the focus of this project. Basalt is a ferromagnesium-rich rock which records the orientation of the Earth's magnetic field as it cools. Paleomagnetic study of basalt flows can therefore be used to reconstruct the movement and timing of displacement along nearby faults if the ages of the basalt flows are known. Carter et al. (1987) conducted paleomagnetic studies on many Joshua Tree basalts and observed that (1) most of the basalt centers have been rotated 41.4 ± 7.7 degrees clockwise, and (2) that the Pinto Well basalt center, dated at 4.5 Ma. (Calzia et al., 1986), is the only basalt flow within the Park that has not been rotated. Carter et al (1987; see also Luyendyk, 1991) developed a model to explain these data, based on the rotation of crustal blocks between left lateral block-bounding faults within and adjacent to Joshua Tree. This model indicates that volcanism in Joshua Tree occurred during left-lateral faulting at and prior to 4.5 Ma, which is the same time frame that the southern San Andreas fault became active (e.g. Crowell, 1975).

Outcrop Geology

Basalts along the Blue Cut fault that have been studied geochemically thus far include the Malapai Hill and Lost Horse centers at the western end of the fault, Mission Well and Eagle Mountain along the eastern segment of the fault, and Pinto Well at the eastern end of the fault. The Smoke Tree and Victory Pass volcanic centers along the Victory Pass fault have also been examined. The basalts at these various centers vary both in texture and structure. Lost Horse and Malapai Hill in the west are composed of massive tilted columnar basalt and contain ultramafic xenoliths, predominantly lherzolite, while the eastern centers are a series of vesicular basalt flows, and do not appear to contain xenoliths.

Mineralogy

The basalts are predominantly dark grey and are aphanitic to porphyritic aphanitic in texture. At four localities they contain seriate ultramafic and mafic xenoliths. Olivine and/or clinopyroxene phenocrysts are enveloped by a clinopyroxene and plagioclase-rich matrix with minor amounts of olivine and opaque oxides. Petrographic analyses revealed two types of olivine crystals; those with evidence of deformation such as undulatory extinction and kink bands and those that lack deformational features (Figure 2). We hypothesize that the deformed olivine crystals are likely xenocrysts that were incorporated into the melt, and the crystals that are not deformed are phenocrysts that crystallized from the melt. To test this idea, olivine phenocrysts and xenocrysts were analyzed using an electron microprobe. Xenocrysts are Fo_{90-70} in composition, compared to Fo_{84-69} for the phenocrysts. The ranges in composition are the result of chemical zoning within each of the crystals, with relatively higher magnesium concentrations in the core (Figure 2). The rims of the xenocrysts were most similar to the composition of the phenocrysts, suggesting that the xenocrysts had in fact been incorporated into the melt as solid phases and later equilibrated to the composition of the melt.

Figure 2a and 2b (facing page). Photomicrographs illustrating the different types of olivine crystals found within Joshua Tree basalts. (a) The distinctly deformed crystal is a xenocryst from Malapai Hill, exhibiting kink bands. (b) These subhedral to euhedral crystals from Pinto Well are typical olivine phenocrysts. The field of view of these images is 3 mm wide.

Figure 2a

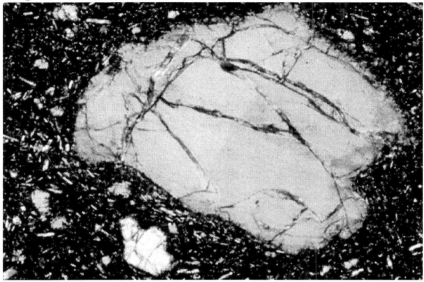

Figure 2b

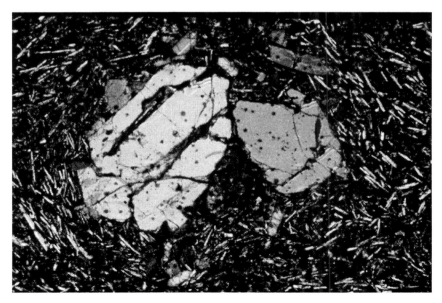

Geochemistry

Major element concentrations for Joshua Tree basalt samples were determined using x-ray fluorescence to investigate possible models for the origin of the rocks. Basalt erupts in three major tectonic settings on Earth, each characterized by a unique chemical signature. Island arc basalt, such as is found in the Aleutian Islands, is distinguished by high Al_2O_3 concentrations; mid-ocean ridge basalts (MORB), such as the lavas on the East Pacific Rise, are recognized by their depletion in alkali elements; and ocean island basalts (OIB), such as the basalts of the Hawaiian islands, are enriched in alkalis and high charge elements.

All of the Joshua Tree Neogene basalt centers studied thus far are comprised of alkali olivine basalt geochemically similar to OIB. These basalts are enriched in high field strength (charge/ionic ratio) elements such as TiO_2 and P_2O_5 compared to nearby alkali basalts of similar age in the Mojave desert, such as the Miocene to Pliocene Cima volcanic field (Farmer and others, 1995) and the Quaternary(?) Amboy crater flows (Glazner and others, 1994). Modern day examples of basalts of similar composition to Joshua Tree alkali olivine basalts are the late alkalic cap lavas of Haleakala volcano on the island of Maui in Hawaii (West and Leeman, 1994).

There are systematic geochemical differences between different Joshua Tree volcanic centers. There are two broad trends that we observe when comparing magnesium and alkali concentrations and there is a distinct difference when comparing longitude and iron concentrations. The Eagle Mountain and Smoke Tree centers show a positive correlation between magnesium and potassium, while the Mission Well, Pinto Mountain, and Victory Pass centers show a more typical fractionation trend with an inverse relationship between the two elements; Malapai Hill plots as the base line of these opposing trends (Figure 3). The western, xenolith-bearing basalts that have been analyzed thus far are relatively enriched in iron compared to basalts further to the east (Figure 4). This trend could also be the result of systematic differences between xenolith-bearing and xenolith-free basalts (see Glazner and Farmer, 1992). The microprobe results for olivine phenocrysts and xenocrysts indicate that the iron enrichment observed in the xenolith-bearing basalts is not the result of xenolith contamination, since the xenoliths and xenocrysts are relatively enriched in magnesium. These geochemical trends suggest that Joshua Tree basalts cannot have been derived from a single parental magma; because of the wide variation in iron and alkali element abundances at similar magnesium contents, more than one magma source was probably involved at the time of eruption of these basalts.

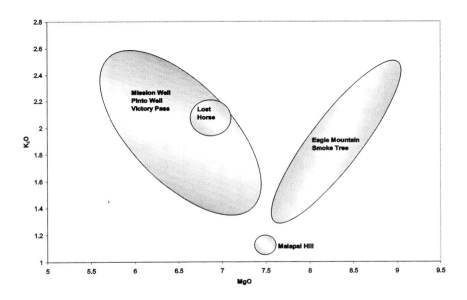

Figure 3. There are two broad trends that are observed when comparing an abundant element such as magnesium to an alkali element such as potassium. Malapai Hill plots as a base line to a typical fractionation trend displayed by basalts from Mission Well, Pinto Well, and Victory Pass, and a positive correlation trend displayed by basalts from the Eagle Mountain and Smoke Tree centers.

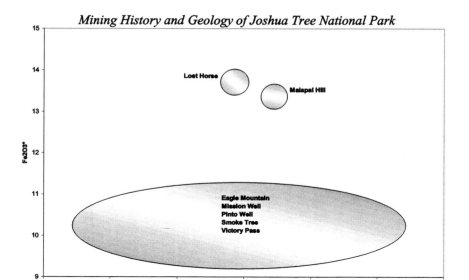

Figure 4. A geographic difference between the eastern and western basalts is observed when comparing iron contents. This observation, when combined with the trends in Figure 3, suggests that more than one magma source was involved at the time of eruption of these basalts.

Modeling

Fractionation models were constructed to determine whether any of the observed geochemical variations noted are the result of olivine, clinopyroxene and/or plagioclase crystallization and fractionation. The fractionation models failed because the crystallization of these minerals alone cannot increase the potassium content enough to explain the trends exhibited by samples from the Mission Well, Pinto Well, and Victory Pass centers, and no combination of the observed mineral phases can cause an increase in magnesium and potassium contents as observed at the Eagle Mountain and Smoke Tree centers. This led us to develop mixing models to assess the potential role of crustal contamination. Using the White Tank granite as a model for the local upper crust, mixing the two components yields the type of inverse trend between magnesium and potassium observed in basalts at several of the Joshua Tree centers. However, when silica concentrations are considered this model quickly fails, because the product of a granite and basalt mixture is not basalt, but instead some intermediate composition lava unlike anything yet observed. Given that crustal contamination is probably not the primary source of the abundance of alkali elements, it is necessary to use isotope geochemistry to begin evaluating the possible influence of the mantle on the composition of Joshua Tree basalt parent magmas.

CONCLUSION

The mineralogy and geochemistry of the Joshua Tree basalts establishes that the volcanic centers here are all characterized by alkali olivine basalt. The basalt was erupted in a tectonic environment dominated by large-scale strike slip faulting. We know that these centers are similar enough that the tectonic setting at the time of eruption was probably the same for all of the volcanoes. However chemical differences seem to preclude the possibility that these volcanoes could have been the product of a single parent magma or erupted from a single magma chamber. To better understand the relationship among the volcanic centers and to facilitate construction of realistic petrologic models, geochronologic and isotopic analyses are underway.

ACKNOWLEDGEMENTS

This manuscript benefited from the technical review of Dr. Dave Kimbrough, San Diego State University Department of Geological Sciences.

REFERENCES

Calzia, J.P., DeWitt, E., and Nakata, J.K., 1986. U-Th-Pb *Age And Initial Strontium Isotopic Ratios Of The Coxcomb Granodiorite, And A K-Ar Date Of Olivine Basalt From The Coxcomb Mountains, Southern California,* Isochron/West 47, 3-8.

Carter, J.N., Luyendyk, B.P., and Terres, R.R., 1987. *Neogene Clockwise Tectonic Rotation Of The Eastern Transverse Ranges, California, Suggested By Paleomagnetic Vectors,* Geological Society of America Bulletin 98, 199-206.

Crowell, J.C., 1975. *The San Andreas Fault In Southern In California, in:* Crowell, J.C. (ed.), San Andreas Fault in Southern California: California Division of Mines and Geology Special Report 118, 7-27.

Farmer, G.L., Glazner, A.F., Wilshire, H.G., Wooden, J.L., Pickthorn, W.J., and Katz, M., 1995. *Origin Of Late Cenozoic Basalts At The Cima Volcanic Field, Mojave Desert, California,* Journal of Geophysical Research 100, 8399-8415.

Glazner, A.F., and Farmer, G.L., 1992. *Production Of Isotopic Variability In Continental Basalts By Cryptic Crustal Contamination,* Science 255, 72-74.

Glazner, A.F., Farmer, G.L., Hughes, W.T., Wooden, J.L., and Pickthorn, W.J., 1991. *Contamination Of Basaltic Magma By Mafic Crust At Amboy And Pisgah Craters, Mojave Desert, California,* Journal of Geophysical Research 96, 13673-13691.

Hope, R.A., 1969. *The Blue Cut Fault, Southeastern California,* U.S. Geological Survey Professional Paper 650-D, D116-D121.

Luyendyk, B.P., 1991. *A Model For Neogene Crustal Rotations, Transtension, And Transpression In Southern California,* Geological Society of America Bulletin 103, 1528-1536.

Powell, R.E., 1981. *Geology Of The Crystalline Basement Complex, Eastern Transverse Ranges, Southern California,* Ph.D. dissertation, California Institute of Technology, 441 pp.

Stull, R.J., and McMillan, K., 1973. *Origin Of Lherzolite Inclusions In The Malapai Hill Basalt, Joshua Tree National Monument, California,* Geological Society of America Bulletin 84, 2343-2350.

West, H.B., and Leeman, W.P., 1994. *The Open-System Geochemical Evolution Of Alkalic Cap Lavas From Haleakala Crater, Hawaii, USA,* Geochimica et Cosmochimica Acta 58, 773-796.

ROAD LOG FOR THE "18-MILE" GEOLOGY TOUR JOSHUA TREE NATIONAL PARK

Margaret R. Eggers, PhD
Eggers Environmental, Inc., Oceanside, California
Phil Farquharson
CG Squared Productions, Point Loma, California

INTRODUCTION

The 18-Mile Geology Tour within the Park wanders through some of the most interesting and varied landscapes and rock types in the Park. The drive is set up as a series of 16 stops (Figure 1), each showing an example of a landscape formation, rock type, geologic feature or mining history. There are excellent examples of the faulting, weathering, erosion and joint sets which have produced these dramatic landscapes.

The Park Service recommends four-wheel drive on the tour due to areas of loose sand and low lying areas which may become soggy at certain times of the year. Recreational vehicles are never recommended along this tour. New restroom facilities are located at the beginning of the tour, at the turnoff from Park Boulevard. Chemical toilets which previously were placed in the valley have been removed. Stops 1 through 9 are located on the 2-way portion of the road. Beyond Stop 9, the road is one-way. Stops are marked by numbered, thin brown stakes that are sometimes hard to see or have been knocked down. The map will help you find the general Stop locations. At the start of the tour, there is an "Iron Ranger" where you can purchase a pamphlet "Geology Tour Road Guide."

ROAD LOG

Mileage shown is the total mileage from the turnoff from Park Boulevard (first number) and the incremental mileage between stops (second number). Since every vehicle is different, your specific mileage may vary. But take your time and enjoy this trip. You can spend between 1 to 2 hours or more depending on how many times you wish to stop, explore, and soak up the geology!

Figure 1. Locations of stops along the 18-Mile Geology Tour

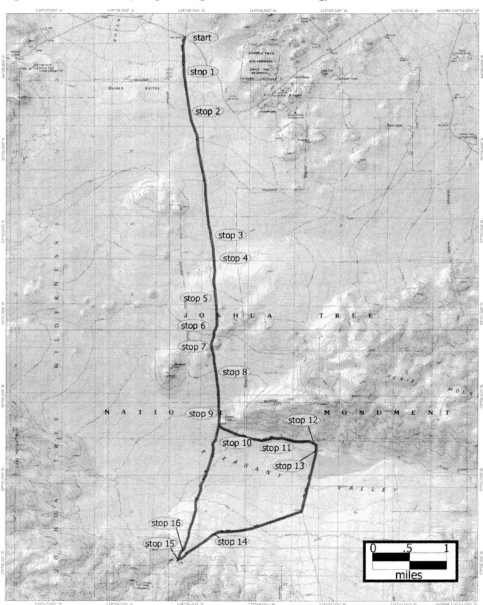

Total Mileage	Incremental Mileage	
0.0	0.0	Turn off from Park Boulevard onto dirt, 2-way road.
0.3	0.3	**STOP 1:** Queen Valley. This picturesque valley is the result of differential erosion (Figure 2). The floor of the valley is composed of White Tank monzogranite, which forms a pediment surface. Granitic inselbergs rising from the valley floor are erosional remnants. The mountains on either side of the valley are composed of the more competent, Pinto Gneiss.

The black hill which appears in the near-center of the valley is an intrusion of basalt, and is named Malapai Hill (see Stop 7). Malapai Hill is a good landmark to keep in sight along the first portion of the Geology Tour, and is also a good point of reference when viewing aerial photographs of this portion of the park.

Figure 2. Malapai Hill and inselbergs of the White Tank in the foreground (photo by D.D. Trent)

1.1	0.8	**STOP 2:** North/South Drainage Divide. Driving up this knoll, one comes to the natural drainage divide for the Park. Overall, Queen Valley is seen topographically as a broad, irregular dome. From here, water drains to the northwest along Quail Springs Wash, and to the southeast through Fried Liver Wash, into Pinto Basin.
1.4	0.3	California Riding and Hiking Trail to Keys View (6.6 miles).
2.7	1.3	**STOP 3:** Wash Crossing. During the summer, intense rainstorms can produce flash flooding along this and other desert washes.
2.9	0.2	**STOP 4:** Old Erosional Level. Outcrops and boulders of the White Tank monzogranite along the left (east) side of the road exhibit a distinct indentation up to seven feet above the current ground surface. Most interpretations believe this indentation indicates a former soil horizon, produced during a time when there was a much moister climate in the region. The wetter climate produced a stable, thick soil horizon and more rapid weathering along the boulder edges. During this time of soil formation, weathering was concentrated along the line of the indentation. As the climate became drier, there was little soil development, the granite was exposed more rapidly, which resulted in less weathering along the sides of the boulders.
3.4	0.5	**STOP 5:** Rock Piles of weathered White Tank monzogranite. Weathering along joint systems in the White Tank monzogranite occurred when these rocks were covered by a thick soil horizon at a time of moister climate. As the climate became more arid, soil development ceased, soils were eroded, and gradually these weathered rock piles

became exposed. See Eggers & Trent (2004), this field guide, for an in-depth discussion of how these landscapes evolve.

4.0   0.6   **STOP 6:** Rock Piles – A Natural Sculpture. Excellent examples of spheroidal weathering. See Eggers & Trent, this field guide, for an in-depth discussion of how these landscapes evolve.

4.5   0.5   **STOP 7:** Malapai Hill. This basalt hill is an intrusion into the surrounding granites. Differential erosion has left the more resistant basalt standing about 400 feet above the valley floor. The basalt intrusion also contains inclusions of the olivine-rich peridotite called lherzolite. These inclusions indicate that the parent magma of this basalt had risen at least 30 miles through the earth before cooling here.

If you hike out to the hill, you will find strong, columnar jointing which juts out at an angle from the northwest side of the hillside (Figure 3). If you don't want to walk the whole way out to Malapai Hill (1.5 miles roundtrip), just wander a little while out into the granites and observe some wonderful erosional formations (Figure 4). See Probst, et al., (2004) in this guide for more information about the chemistry of the Malapai Hill basalt.

Figure 3. Basalts of Malapai Hill showing strong columnar jointing and tilted orientation (photo by D.D. Trent).

Figure 4. Erosional forms in granites along the way to Malapai Hill (Stop 7, photo by Peter S. Gorwin).

| 4.9 | 0.4 | **STOP 8:** Alluvial Fans and Bajada Development. Looking across the valley, you can see that alluvial fans which developed along the base of the mountains have grown to the extent that they have coalesced into an apron of erosional debris which skirts the base of the mountains. This apron of rock debris along the base of the mountains is known as a bajada. |

| 5.2 | 0.3 | **STOP 9:** Squaw Tank. Huge outcrops of the White Tank monzogranite are present at this turnout. If you wander around, you will find excellent examples of the honeycomb weathering known as tafoni. There are also good exposures of granitic or pegmatitic dikes which cut through the monzogranite.

A bedrock mortar, resulting from grinding of grains by local Indians, is located at the base of the rock apron of the rock pile adjacent to the parking area. Cattlemen constructed a small concrete dam in a natural catchment just beyond the parking area. |

| 5.3 | 0.1 | Intersection with the Pleasant Valley loop portion of the tour. Stay to the left. From this point, the road is one-way only. The western portion of the Hexie Mountains are to your left (north) as you wind your way down the valley. |

| 5.4 | 0.1 | **STOP 10:** Pleasant Valley. From this view you look straight down the north side of Pleasant Valley. The Blue Cut fault runs right along the base of the hills, producing a sharp topographic contact with the valley floor (Figure 5). It is named for the blue granodiorite exposed to the SW, marking the main fault branch. |

Figure 5. Looking east down Pleasant Valley (photo by P. Farquharson, 2004).

6.1   0.7   **STOP 11:** Debris Flows. Rock debris from the hill sides has accumulated along the north side of the road. Occasionally during heavy rains, debris flows have oozed in a slow-moving mass downward. Individual, newer flows can be seen along the roadside. Rock debris mantling the slopes in this area has developed a strong, dark rock varnish. Newer flows are easy to spot due to the fresh, gray, newly exposed rock surfaces that had not previously been exposed or developed the dark rock varnish.

6.6   0.5   Mine workings on the left (north) side of the road, on the hillside.

6.7   0.1   **STOP 12:** Mining in Pleasant Valley – The Gold Coin Mine. The mountains to the left (north) and in front of you (east) are riddled with mine workings, shafts and tunnels dug by prospectors in the late 1800s and early 1900s. The search for gold, silver and other precious metals motivated many a miner to try to turn a profit in the Joshua Tree area.

Unfortunately, most mines were not profitable. The Lost Horse Mine was one of the few exceptions, due to higher grade ore. The Lost Horse Mine was located on Lost Horse Mountain, west of Malapai Hill. In this area, it is thought that fluids migrated along the fault zone, producing the economic mineral deposits.

6.8   0.1   **STOP 13:** Dry Lake. When the climate was wetter, a lake periodically formed within Pleasant Valley. The playa you are driving across was formed from the deposition of silts and clays to a depth of several hundred feet. Like many desert playas many salts were also deposited as the lake dried up, although those salt deposits are not as obvious here. However, the presence of a high salt content is revealed by the occurrence of salt-tolerant vegetation.

7.8   1.0   **Valley view:** The view north from this area is wonderful. Clearly visible is the Blue Cut fault, the contact between the dark Pinto Gneiss and the light monzogranite, Malapai Hill, mine workings and the valley floor. Take your time here as the road along this section is rocky with Pinto Gneiss debris.

9.3   1.5   **STOP 14:** Pinto Gneiss. This is a great exposure of the Pinto Gneiss. This banded and folded formation is approximately 1.7 billion years old. Note the bright splotches of color provided by the growth of lichens on the surface of gneiss outcrops.

9.4   0.1   **STOP 15:** Pinyon Well Junction. This old road, now closed to vehicular traffic, leads up Pushwalla Canyon to the site of a well that provided water for ore processing and livestock. Around 1920, a two-stamp ore processing mill was located here due to the convenient water source. This same mill was later moved to the Wall Street Mill site in Queen Valley (see Dee Trent's mining article in this guide).

9.4   **0.0**   **STOP 16:** Panoramic View. Stop and enjoy this wonderful view of Pleasant Valley and Queen Valley (Figure 6). You can clearly see the road you traveled down into Pleasant Valley.

Figure 6. Looking northeast from Stop 16. The contact between the white monzogranite and the dark Pinto Gneiss, near Stop 10, can be seen clearly in the distance (photo by P. Farquharson, 2004).

9.5   **0.1**   End of the loop. Re-join the main road. Remember traffic is two-way along this portion of the tour.

16.6   **7.2**   Return to Park Boulevard intersection.

REFERENCES AND ADDITIONAL READING

Cates, Robert B., 1995. *Joshua Tree National Park,* printed by Live Oak Press, Chatsworth, California.

Eggers, Margaret R., and Trent, D.D., 2004. *Overview of the Landscape and Geology of Joshua Tree National Park, in*: Mining History and Geology of Joshua Tree National Park, San Diego Association of Geologists Annual Field Guide, 2004, Margaret R. Eggers, editor. Sunbelt Publishing, San Diego California.

Joshua Tree National Park Association, undated. *Geology Tour Road Guide: An 18 Mile Motor Tour,* (pamphlet available for purchase from "Iron Ranger" at start of tour).

Trent, D.D., and Hazlett, Richard W., 2002. *Joshua Tree National Park Geology*, published by the Joshua Tree National Park Association.

Split Rock (Photo by John D. Clark, 2004).

ROAD LOG FOR PARK BOULEVARD FROM WEST ENTRANCE STATION TO COTTONWOOD SPRINGS

D.D. "Dee" Trent, PhD, *Professor Emeritus,*
Citrus College, Glendora, CA 91741
Rick Hazlett, PhD, *Director of the Environmental Analysis*
Program and Professor, Department of Geology,
Pomona College, Claremont, CA 91711

ROAD LOG

This road log follows the main park road, known as "Park Boulevard" from the western park entrance southeast of the town of Joshua Tree. Mileage shown is the total mileage from the West Entrance Station (first number) and the incremental mileage between stops (second number).

Total Mileage	Incremental Mileage	
0.0	**0.0**	**Joshua Tree National Park West Entrance Station:** West Entrance Station. Set your vehicle odometer to "zero." Hillsides on both sides of the highway show jointed granite knobs.
2.4	**2.4**	**Exhibit Pullout:** Partially boulder-mantled slopes. The surrounding hills consist of Mesozoic granitic rocks, probably the White Tank monzogranite. They reveal good examples of boulder-mantled slopes (Oberlander, 1972; Trent and Hazlett, 2001).
2.6	**0.2**	Pediment surface can be viewed on the north side of the road. Vegetation here is a Joshua Tree grassland, which includes some junipers, typical of much of the higher elevations in the Mojave Desert.
3.5	**0.9**	Excellent example of boulder mantled slope immediately north (left) of road. To south (right), an inselberg features both tectonic (conjugate) and lift joints. The reason for the boulder mantles here is that this is a young landscape. There has been insufficient time since soils have washed off the nearby slopes for chemical weathering and erosional processes to break down the boulder

mantles. In the distance, 1/4 mile to the SW, the mountains are held up by 1.7-1.4 billion year old Paleoproterozoic gneiss—rock as old as about a third of the age of the Earth. The gneissic mountains are darker, generally smoother than the mountains composed of light-colored granite.

5.7   2.2   Quail Springs Pullout: Granite inselberg with conspicuous flat-lying pegmatitic veins. Sub-horizontal lift joints show well. From Quail Springs eastward for the next 2 miles the highway passes many inselbergs that reveal numerous intersecting lift and tectonic joints with well-rounded boulders formed by subsoil weathering along these joint surfaces during times of more humid climate.

Figure 1. White Tank monzogranite outcrops in Hidden Valley. Two joint sets are strongly expressed in this outcrop (photo by D.D. Trent).

8.7   3.0   Entrance to Hidden Valley Nature Trail parking area: This one time cattle rustler's hideout offers excellent examples of inselbergs with lift and tectonic joints (Figure 1). Notice the absence of smaller spheroidal boulders. Most of this area has relatively little in the way of boulder mantling, with only a few of the larger spheroidal boulders remaining. The reason is that this region has been subjected to a considerably longer period of weathering and erosion than have the inselbergs and rocky knobs seen previously at lower elevations.

The Hidden Valley Nature Trail takes advantage of the joint system in the granite. Where the trail follows a narrow slot between high granite walls, erosion has preferentially removed rock along steep, closely spaced tectonic joints. Many of the steps in the trail, such as those that are especially obvious near the wooden footbridge, are surfaces broken out along near horizontal lift joints.

11.4   2.7   Junction with Keys View Road and Cap Rock

12.0   0.6   To the east (right), well exposed pediment between the road and the foot of Ryan Mountain and the sharp contact of the Mesozoic White Tank monzogranite with the Proterozoic gneiss about midway up the slope of Ryan Mountain (Figure 2). Note the abrupt break in slope, one of the characteristic geomorphic landforms of arid regions, where the pediment ends and the steep mountain slope begins.

Figure 2. White Tank monzogranite intruding into the darker Proterozoic gneiss (photo by D.D. Trent)

12.5 0.5 To the east (right), the massive, bold, unjointed knobs of the White Tank pluton in the lower part of Ryan Mountain are in marked contrast to the highly jointed darker Queen Mountain pluton in the hill slopes to the north (ahead).

12.9 0.4 Sheep Pass Group Campground

13.6 0.7 Crossing Sheep Pass and entering Queen Valley. Three different types of granite have intruded the gneiss in this area with each granite weathering in a characteristically different manner. Note the many very large, i.e., very old, Joshua Trees bordering the road.

15.4 1.8 Junction with Geology Tour Road: Highway is just beginning to cross a pediment. For a road log to the Geology Tour Road, see Eggers and Farquharson (2004), this guide.

15.9 0.5 Scattered, low relief exposures of granite north (left) and south (right) of road show this area to be a pediment.

16.0 0.1 Jumbo Rocks Campground

16.4 0.4 Weathered pegmatite veins in pediment to the south (right). The veins have weathered into small cobbles and boulders that appear to be arranged in lines on the bedrock pediment surface.

16.6 0.2 Skull Rock Turnout: This formation has spectacular examples of tafoni in the weathered granites. Large cavernous tafoni form the "eye" sockets in Skull Rock (see Figure 6, Eggers and Trent, 2004, this guide).

16.9 0.3 Exhibit and turnout: north side of the road.

17.3 0.4 Junction with Split Rock Road

17.8 0.5 Pullout on south (right) side of road. Pinto Mountains to the east (straight ahead) are composed mostly of 1.7-1.4 billion year old gneiss.

19.3 1.5 Intersection: Turn right to continue to Cottonwood Campground. The North Entrance of the Park and Twentynine Palms is to the north (left turn). In the distance to the north you can see several outcroppings of irregular granite intrusions that protrude through the gneiss.

20.1   0.8   Road to Belle Campground: Granite inselbergs lack boulder mantles indicating a long period of erosion and weathering. Good view of the Pinto Basin can be seen ahead.

20.7   0.6   Transition from a granitic terrain, to a terrain underlain by dark metamorphic rocks (gneiss).

21.7   1.0   Road to White Tank Campground: The one-mile long Arch Rock Geology Nature Trail in this campground is a worthwhile, well-labeled and informative side trip.

22.1   0.4   The road begins to descend into the Pinto Basin by following a canyon that has been cut into the pediment we have been crossing. It was eroded by running water during earlier times when the climate was cooler and more humid than today.

23.1   1.0   Crossing an arroyo with numerous smoke trees. Beginning about here for approximately the next five miles, the highway crosses an ecological transition zone between the Mojave Desert, behind you and to the north, and the Colorado Desert, ahead of you and to the south. The Colorado Desert is a lower, hotter, and drier desert than the Mojave. Plant types reflect this difference. The greenish-gray smoke trees that you see near the road, for example, are indigenous to the Colorado Desert. Smoke trees are largely restricted to arroyos. It is here that flash floods in the summer cause the seeds of the smoke trees to be abraded, a processes that is necessary before the seeds can germinate.

24.0   0.9   Exhibit pullout: Gneiss with minor veins of granite or quartz occur on both sides of the highway.

25.6   1.6   Good view of the Pinto Basin where it is easy to appreciate its origin as a down-dropped basin along normal faults along its northern and southern boundaries.

26.7 1.1 Sign stating, "No Stopping the Next Mile." To the south (right) are fault scarps along the base of the Hexie Mountains (Figure 3). These scarps mark the position of two parallel branches of the east-west trending Blue Cut fault, and their freshness indicates recent displacement along this fault. To north (left), along the lower flanks of the Pinto Mountains, are alluvial fans.

27.1 0.4 Cholla Cactus Garden: The floor of Pinto Basin in this area is rich in picturesque cholla cactus. There is a self-guided nature trail you may enjoy – but be careful, Cholla spines are very painful!

Figure 3.   Step-like fault scarps cut the base of the Hexie Mountains.   These scarps mark two branches of the Blue Cut fault, viewed from Park Boulevard near the Cholla Cactus garden (photo by D.D. Trent).

28.6 1.5 Ocotillo Cactus Garden: To the south (right), an alluvial fan extends northward from the base of the Hexie Mountains that buries much of the Blue Cut fault. East of the fan, a linear break in the foothills of the Hexie Mountains to the southeast marks the continuation of the Blue Cut fault.

At this point, the Park road has left the Mojave Desert and entered the Colorado Desert. The particular cactus species and other plants here are typical of the Colorado Desert biome.

29.6 1.0 Far in the distance, about 25 miles to the east (left), are the granitic Coxcomb Mountains which mark the eastern boundary of the Park.

30.3 0.7 Fried Liver Wash

32.8 2.5 To the north, at the western end of the Pinto Basin, are low hills noted as sand dunes on the hiking maps of Joshua Tree National Park. True, there are some thin sand sheets among those hills, but in reality they are uplifted sediments that accumulated on the bottom of the Pleistocene lake that occupied Pinto Basin. The fluvial/lacustrine beds have been tilted upward by movement on one strand of the Blue Cut fault.

33.4 0.6 The Pinto Mountains to the north (left), with Pinto Mountain (elev. 3,983 ft.) the highest peak, are composed generally of Barth et al. (2004, this volume) Mesoproterozoic age sedimentary rocks that they informally name the Pinto Mountain Group. This supracrustal suite of rocks consists of white quartzite, laminated Fe-rich rocks, dark pelitic rocks, and carbonates which combine to give the range its distinctive color pattern; hence the name "Pinto Mountains." The suite has been metamorphosed to the upper greenschist-lower amphibolite facies and detrital zircon geochronology suggests the suite was deposited between about 1630 Ma and 1450 Ma. Thus, their age corresponds to that of the Belt and Wernecki supergroups of southern Canada and/or the upper part of the Tonto Basin supergroup of Arizona (Barth, et al., 2004, this paper).

33.6 0.2 To the south (right) an alluvial fan formed on the lower flanks of the Hexie Mountains has been cut by a fan head trench.

36.6 3.0 Low relief boulder mantled slopes to the south (right). Their dark color is misleading, the rock being a light-colored granite but the color due to a heavy coating of rock varnish. Does this imply a longer period of weathering? If so, why do so many smaller spheroidal boulders still exist? Because dust is necessary for rock varnish to form, bacteria metabolically fixing the dust to form the varnish, perhaps this area has received more dust in a shorter time span than elsewhere in the Park, the nearby Pinto Basin washes perhaps being the source of abundant dust.

42.1 5.5 Smoke Tree Wash

42.5 0.4 Exhibit on east (left) side of road. In the distance to the west (right) is the southern slope of the Hexie Mountains with a prominent Mesozoic granodiorite mass intruding the darker Proterozoic gneiss (Figure 4). Varnish and weathering of the boulder mantles for the next few miles is more similar to that seen in the Hidden Valley region than that reported in the Pinto Basin.

Figure 4. Lighter Mesozoic granodiorite intrudes the darker Proterozoic gneiss (photo by Phil Farquharson, 2004).

46.3 3.8 To east (left) is a well-developed pediment eroded on granite.

46.7 0.4 Cottonwood Visitor Center: The pediment, sloping upward toward the mountains in the distance, underlies the surface to the east. Turn left from the Cottonwood Visitor Center onto the paved road leading to Cottonwood Spring.

47.0 0.3 Cottonwood Spring: The spring is reached via a short path from the parking area. Like virtually all springs in this desert, it lies along a fault. Remains of an early-day *arrastra*, a simple but effective apparatus used to mill ore in order to recover gold (Figure 5), is located a few feet south of Cottonwood Spring. If you continue southward toward Interstate 10, the road follows Cottonwood Wash that separates the Cottonwood Mountains on the west from the Eagle Mountains on the east. The many Smoke trees along the wash attest to the frequency of flash floods during the rainy season. Palo Verde trees and ocotillo are also common, all of these plants requiring some summer water.

50.6 3.6 Parking turnout on east (left) side of road. To the east (left) across the wash is a spectacular contact of granite intruding gneiss. Much of the contact is marked by a textbook example of an intrusion breccia — a mass of older rock broken up and set within a younger igneous intrusion.

Figure 5. How the arrastra at Cottonwood Spring might have looked when it was used to mill ore (drawing by D.D. Trent)

53.1 2.5 **Leaving Joshua Tree National Park:** To the south (ahead), is a spectacular bajada draping the base of the northern flank of the Orocopia Mountains. As you emerge from Cottonwood Canyon and leave the Park, the southern border of the Cottonwood Mountains marks the approximate position of the Chiriaco fault, a left lateral, east-west fault with about 11 km of offset (Powell, 1982). Ahead, the Orocopia Mountains consist of Miller's (1944) Orocopia Schist. The western end of the highest part of the eastern Orocopias, visible south of Chiriaco Summit, is underlain by Mesozoic plutons (Powell, 1982).

REFERENCES

Barth, Andrew P., Wooden, Joseph L., and Jarvis, Janet L., 2004. *Crust Formation And Evolution In Southern California: Field And Geochronologic Perspectives From Joshua Tree National Park, in*: Mining History and Geology of Joshua Tree National Park, San Diego Association of Geologists Annual Field Guide, 2004, Margaret R. Eggers, editor. Sunbelt Publishing, San Diego California.

Miller, W.J., 1944. *Geology of the Palm Springs-Blythe strip, Riverside County, California.* California Journal of Mines, vol. 40, 11-72.

Oberlander, T.M., 1972. *Morphogenesis of granite boulder slopes in the Mojave Desert, California.* Journal of Geology, vol. 80, no. 12, 1-20.

Powell, Robert E., 1982. *Crystalline basement terranes in the southeastern Transverse Ranges, California. in:* Cooper, J.D., editor, Geologic Excursions in the Transverse Ranges: Geological Society of America Cordilleran Section, 78th annual meeting, Anaheim, California, 109-136.

Trent, D.D., and Hazlett, Richard W., 2002. *Joshua Tree National Park Geology*, published by the Joshua Tree National Park Association.

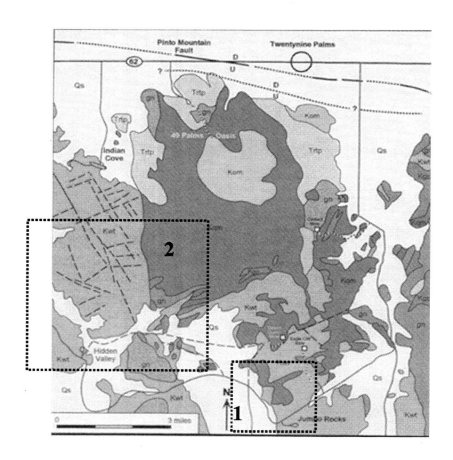

Approximate area covered by aerial photographs 1 and 2.
See color geologic map on inside front cover of this guide.

SELECTED AERIAL VIEWS OF JOSHUA TREE NATIONAL PARK

Margaret R. Eggers, PhD, *Eggers Environmental, Inc.,*
Oceanside, CA 92054
Woodrow L. Higdon, *Geo-Tech-Imagery,*
Oceanside, CA *www.geo-tech-imagery.com*

Editor's note: Note that we have oriented the explanations for these aerial views so that you can read them while you look at the photos in the proper orientation!

Photos 1 and 2 cover areas included in the map "Geology of the Central Part of Joshua Tree National Park," provided in color on the inside cover of this Guide. An index map showing the approximate location of Photos 1 and 2 is shown on the facing page.

Photo 3 covers an area of Pleasant Valley and Malapai Hill along the 18-Mile Geology Tour. Compare this photo with the map showing the 16 stops on the Geology Tour, shown on page 90 of this guide.

Composite Aerial Photograph 1, facing page:

This aerial photograph shows Park Boulevard (since it's an asphalt road, it is visible here as a black line) where it runs through the Jumbo Rocks area in the Mojave Desert section of the Park. The first portion of the "18-Mile Geology Tour" is visible as a dirt road (light colored line) extending south from Park Boulevard on the left side of the photograph. Note the strong lineations visible in the light colored monzogranite outcrops. The lighter, heavily jointed rocks covering most of the eastern portion of the view are those of the White Tank monzogranite. This area correlates with the southeast portion of the "Geology of the Central Part of Joshua Tree National Park" map, located on the inside front cover of this guide.

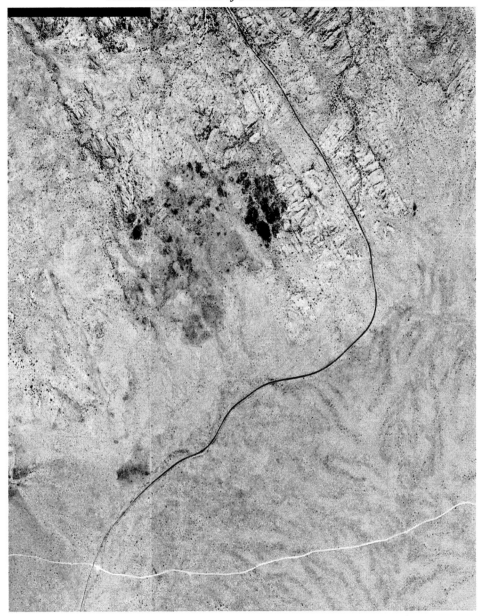

< < < N O R T H

Composite Aerial Photograph 2, facing page:

This aerial photograph shows the contact between the strongly jointed White Tank monzogranite on the left (west) portion of this view, and the Queen Mountain monzogranite to the right (east). The northern portion of Queen Valley is present in the bottom right (southeast) corner of this view.

Note the strong lineations trending north-northwest through the White Tank monzogranite. These strongly expressed joint patterns are key to the origin of the dramatic rock piles, scenic outcrops, and the great rock climbing, that have earned this area the name "Wonderland of Rocks." The Queen Mountain monzogranite on the eastern portion of this view does not show the strong lineations, and is also more vegetated. Small outcrops of darker Proterozoic gneiss can also be seen in this view.

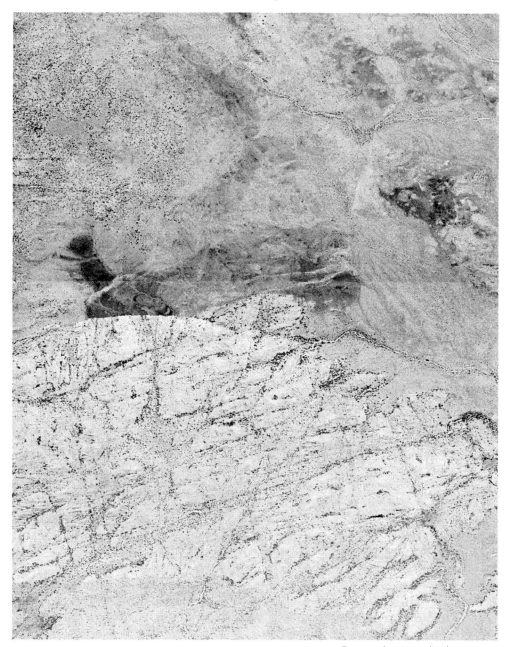

‹ ‹ ‹ N O R T H

Composite Aerial Photograph 3, facing page:
This aerial photograph shows a portion of Pleasant Valley, in the area of the loop on the 18-Mile Geology Tour. The southern portion of Malapai Hill can be seen in the upper left (northwest) corner of this view. The trace of the Blue Cut fault runs nearly east-west across the photograph, along the base of a portion of the Hexie Mountains. Compare this photograph to the illustration showing the 16 stops on the 18-Mile Geology Tour, on page 90 of this Guide.

< < < N O R T H